Beyond Global Warming

BEYOND GLOBAL WARMING

How Numerical Models Revealed the Secrets of Climate Change

Syukuro Manabe and Anthony J. Broccoli

PRINCETON UNIVERSITY PRESS

Princeton and Oxford

Published by Princeton University Press
41 William Street, Princeton, New Jersey 08540
6 Oxford Street, Woodstock, Oxfordshire OX20 1TR

press.princeton.edu

ISBN 978-0-691-05886-3
ISBN (e-book) 978-0-691-18516-3

British Library Cataloging-in-Publication Data is available

Editorial: Jessica Yao and Arthur Werneck
Production Editorial: Jenny Wolkowicki
Text and jacket design: Chris Ferrante
Production: Jacqueline Poirier
Publicity: Matthew Taylor and Katie Lewis
Copyeditor: Maia Vaswani

This book has been composed in Minion Pro, Gotham, and Alternate Gothic

Printed on acid-free paper. ∞

Printed in the United States of America

10 9 8 7 6 5 4 3 2

CONTENTS

Figures

Tables

Plates (following p. 94)

There is no doubt that the composition of the atmosphere and the Earth's climate have changed since the industrial revolution, with human activities as the predominant cause. The atmospheric concentration of carbon dioxide has increased by more than 40% since the preindustrial era, primarily from the combustion of fossil fuels for the production of energy. The global mean surface temperature, which has been relatively stable over 1000 years, has already increased by about 1°C since the preindustrial era. If these energy production activities do not shift markedly, these changes will inevitably continue. The global mean temperature is projected to increase by an additional 2°C–3°C during the twenty-first century, with land areas warming significantly more than oceans and the Arctic warming significantly more than the tropics.

The availability of water is also likely to change over the continents. Water will probably be more plentiful in already water-rich regions, increasing the rate of river discharge and frequency of floods. In contrast, water stress will increase in the subtropics and other water-poor regions that are already relatively dry, increasing the frequency of drought. Observations suggest that the frequencies of both floods and droughts have been increasing. Unless dramatic reductions of greenhouse gas emissions are achieved, global warming is likely to exert far-reaching impacts upon human society and the ecosystems of our planet during the remainder of this century and for many centuries to come.

Climate models are the most powerful tools for predicting human-induced global warming. They are based upon the laws of physics and have evolved from the models used for numerical weather prediction. Exploiting the vast computational resources of some of the world's most powerful supercomputers, climate models have been used to make predictions of future climate change and its impacts, providing valuable information for policymakers. Climate models have been useful not only for predicting climate change but also for understanding it. Serving as "virtual laboratories" of the coupled atmosphere-ocean-land system, they can be used for performing controlled experiments that have proven very effective for systematically elucidating the physical mechanisms involved in climate change.

The primary title of this book, *Beyond Global Warming*, reflects our strong belief that the greatest value of climate models is not just their utility for making predictions, but also their ability to provide a deeper understanding of how the climate system works. Starting from the pioneering study conducted by Arrhenius more than 100 years ago, this book presents a history of the use of models in studies of climate change. Based upon the analysis of many numerical experiments performed with a hierarchy of climate models of increasing complexity, we seek to elucidate the basic physical processes that control not only global warming but also the changes in climate of the geologic past. It is not our intention, however, to present a comprehensive survey of the literature on climate dynamics and climate change. Instead, we would like to focus on studies in which Manabe was a participant and those that influenced his thinking. We hope to describe the scientific journey that allowed him to develop a better understanding of the processes that underlie climate change. He was accompanied for parts of this journey by Broccoli, who was likewise influenced and informed by the studies described in this volume.

This book has evolved from the lecture notes of a graduate course that Manabe taught in the Program in Atmospheric and Oceanic Sciences at Princeton University. The book may be useful as a reference text for graduate and advanced undergraduate courses in climate dynamics and climate change, but also in other disciplines that involve the environment, ecology, energy, water resources, and agriculture. But, most of all, we hope that this book will be useful for those who are curious about how and why the climate has changed in the past and how it will change in the future.

ACKNOWLEDGMENTS

We would like to dedicate this book to the late Joseph Smagorinsky, the founding director of the Geophysical Fluid Dynamics Laboratory (GFDL) of the National Oceanic and Atmospheric Administration of the United States, where we conducted almost all of our studies mentioned in this book. His superb leadership, inspiration, and professional influence enabled us to construct climate models and conduct countless numerical experiments that explored the physical mechanisms of past, present, and future climate change.

We thank Kirk Bryan, who pioneered the development of general circulation models of the ocean. Working with him to develop a coupled ocean-atmosphere model and to explore the role of the ocean in climate change has been a great privilege and a pleasure.

The publication of this book would not have been possible without the encouragement and wholehearted support of the current director of GFDL, Dr. V. Ramaswamy, and the former director of the Atmospheric and Oceanic Sciences Program of Princeton University, Professor Jorge Sarmiento, who have so generously made available the resources of their institutions for this undertaking.

We thank Dennis Hartman, Matthew Huber, and Raymond Pierrehumbert, who read a draft of this book and provided comments that have helped us improve the manuscript. We are also grateful for the efforts of the staff of Princeton University Press, who worked with us to bring this project to completion.

Finally, we thank our wives and life partners, Nobuko Manabe and Carol Broccoli, for their unfailing encouragement during the preparation of this book. We could not have completed it without their patient and unwavering support.

Beyond Global Warming

CHAPTER 1

Introduction

The global surface temperature has increased gradually since the turn of the twentieth century. This is evident in figure 1.1, which shows the time series of the global mean surface temperature anomaly (relative to a 1961–90 baseline) since the middle of the nineteenth century. Although the temperature fluctuates over interannual, decadal, and interdecadal time scales, it has increased gradually during the past 100 years with a relatively large increase during the past several decades. For the period prior to the middle of the nineteenth century, many attempts have been made to reconstruct the large-scale trends of surface temperature from natural archives of climate-related proxies, using data sets that include more than 1000 tree-ring, ice-core, coral, sediment, and other assorted proxy records from ocean and land throughout the globe (Jansen et al., 2007). As one example of such efforts, figure 1.2 depicts the time series of Northern Hemisphere mean surface temperature reconstructed by Mann et al. (2008, 2009) for the past 1500 years. According to this reconstruction, surface temperature was relatively low between 1450 and 1700 (i.e., the Little Ice Age) but was relatively high prior to 1100, during the Medieval Climate Anomaly. These results suggest, however, that the warmth of the past half-century is quite unusual during at least the past 1500 years.

As stated in the *Fifth Assessment Report of the Intergovernmental Panel on Climate Change* (IPCC, 2013b), "it is extremely likely that human influence has been the dominant cause of the observed warming since the mid-20th century" (17). The report further concludes that a majority of the observed warming can be attributed to the anthropogenic increase in the concentrations of greenhouse gases such as carbon dioxide, methane, and nitrous oxide. According to figure 1.3, which shows the temporal variation in carbon dioxide (CO_2) concentration over the past 1200 years, the level of atmospheric CO_2 fluctuated by around 280 ppmv (parts per million by volume) until the end of the eighteenth century,

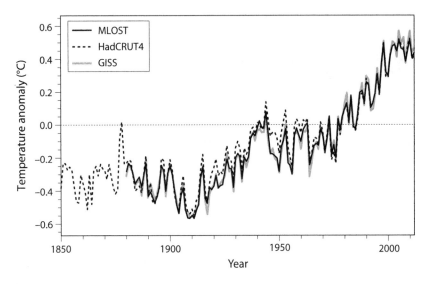

FIGURE 1.1 Annual global mean surface temperature anomalies (relative to the 1961–90 reference period mean) from the latest versions of the three combined land-surface-air temperature (LSAT) and sea surface temperature (SST) data sets (HadCRUT4, GISS, and NCDC MLOST). For the identification of the three institutions indicated by acronyms, see Intergovernmental Panel on Climate Change (IPCC) (2013a). From Hartmann et al. (2013).

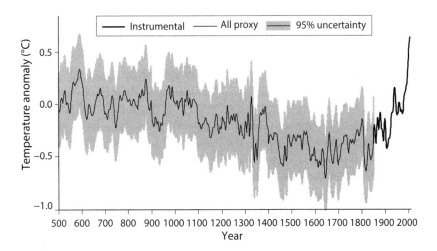

FIGURE 1.2 Time series of Northern Hemisphere mean surface temperature. The thin line indicates the temporal variation of reconstructed decadal surface temperature anomaly (°C), averaged over the entire Northern Hemisphere, during the past 1500 years. Anomalies are defined relative to the 1961–90 reference period mean; shading indicates 95% confidence intervals. For the recent period, anomalies obtained by thermometers are indicated by the thick line. From Mann et al. (2008, 2009).

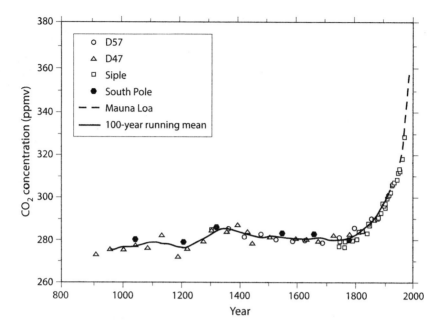

FIGURE 1.3 Temporal variation of CO_2 concentration in air over the past 1100 years, from Antarctic ice-core records (D57, D47, Siple, South Pole) and (since 1958) data from the Mauna Loa measurement site. The former are based upon the analysis of air bubbles in the Antarctic ice sheet, and the latter were obtained from instrumental observations. The smooth curve is based on a 100-year running mean. From Schimel et al. (1995).

when it began to increase gradually. This increase accelerated during the twentieth century, when temperature also increased as indicated in figure 1.1. Other greenhouse gases such as methane and nitrous oxide have also increased in a qualitatively similar manner during the same period. Although greenhouse gases are minor constituents of the atmosphere (table 1.1), they strongly absorb and emit infrared radiation, exerting the so-called "greenhouse effect" that helps to maintain a warm and habitable climate at the Earth's surface.

In the remainder of this chapter, we describe the mechanism by which these gases affect the temperature of the Earth's surface by modifying the upward flux of infrared radiation emitted by it. We then discuss how the greenhouse effect increases in magnitude as the concentrations of these gases increase, warming not only the Earth's surface but also the entire troposphere owing to the upward transport of heat by turbulence and convection.

TABLE 1.1 *Composition of the air*

Constituent	Approximate % by weight
Nitrogen (N_2)	75.3
Oxygen (O_2)	23.1
Argon (Ar)	1.3
Water vapor (H_2O)*	~0.25
Carbon dioxide (CO_2)*	0.046
Carbon monoxide (CO)	~1×10^{-5}
Neon (Ne)	1.25×10^{-3}
Helium (He)	7.2×10^{-5}
Methane (CH_4)*	7.3×10^{-5}
Krypton (Kr)	3.3×10^{-4}
Nitrous oxide (N_2O)*	7.6×10^{-5}
Hydrogen (H_2)	3.5×10^{-6}
Ozone (O_3)*	~3×10^{-6}

*Greenhouse gases.

The Greenhouse Effect

The Earth exchanges energy with its surroundings via the transfer of electromagnetic radiation. Thus the heat balance of the Earth is determined by the heat gain due to the absorption of incoming solar radiation with relatively short wavelengths (~0.4–1 μm) and the heat loss due to outgoing terrestrial radiation with relatively long wavelengths (~4–30 μm), as illustrated schematically in figure 1.4. In a hypothetical situation in which the energy output of the Sun and the composition of the Earth's atmosphere are unchanged, the net incoming solar radiation and outgoing terrestrial radiation averaged over the entire globe and over a sufficiently long period of time would be exactly equal to each other. This is because the Earth as a whole seeks the temperature that satisfies the requirement of radiative heat balance. If the temperature of the Earth is too high, for example, the heat loss due to outgoing terrestrial radiation is larger than the heat gain due to net incoming solar radiation, thereby reducing the temperature of the planet as a whole. On the other hand, if the temperature is too low, the reverse is the case, raising the temperature of the planet. In the long run, the Earth's temperature is maintained such that the net incoming solar radiation and outgoing terrestrial radiation at the top of the atmosphere are in balance.

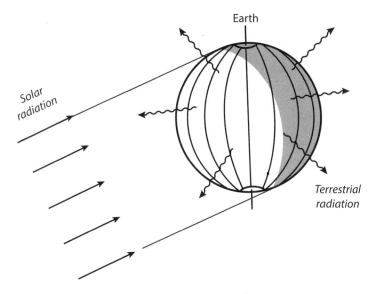

FIGURE 1.4 Schematic diagram illustrating the radiative heat budget of the Earth.

Averaged over the entire globe, the incoming solar radiation at the top of the atmosphere is 341.3 W m^{-2} (Trenberth et al., 2009), of which 101.9 W m^{-2}, or about 30%, is reflected back into space by the Earth's surface, cloud, aerosols, and air molecules. The remaining 70% is absorbed, mostly by the surface, implying that the net incoming solar radiation at the top of the atmosphere is 239.4 W m^{-2}, which is slightly larger than the value of 238.5 W m^{-2} for outgoing terrestrial radiation that has been obtained based upon satellite observations (Loeb et al., 2009; Trenberth et al., 2009). The radiative imbalance of 0.9 W m^{-2} is consistent with the planetary warming that is currently under way. Assuming that the Earth-atmosphere system radiates as a blackbody according to the Stefan-Boltzmann law (see box Blackbody Radiation and Kirchhoff's Law on p. 8), one can compute approximately the effective emission temperature of the planet. The temperature thus obtained is −18.7°C, which is about 33°C lower than the Earth's global mean surface temperature of +14.5°C.

Because the Earth's surface radiates almost as a blackbody, as noted above, one can also use the Stefan-Boltzmann law to estimate approximately the upward flux of radiation emitted by the Earth's surface. The flux thus obtained is 389 W m^{-2} and is much larger than the 238.5 W m^{-2} of outgoing terrestrial radiation emitted from the top of the atmosphere. This implies that the atmosphere traps ~151 W m^{-2} of the upward flux of radiation emitted by the Earth's surface before it reaches the top of the atmosphere.

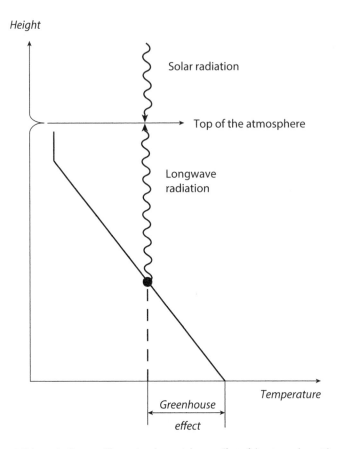

Height

Solar radiation

Top of the atmosphere

Longwave
radiation

Temperature

Greenhouse

effect

FIGURE 1.5 Schematic diagram illustrating the greenhouse effect of the atmosphere. The slanted solid line indicates the vertical temperature profile of the troposphere. The vertical line segment at the top of the slanted line indicates the almost isothermal temperature profile of the stratosphere. The dot on the slanted line indicates the average height of the layer of emission for the upward flux of outgoing terrestrial radiation from the top of the atmosphere.

In short, the atmospheric greenhouse effect prevents approximately 39% of the outgoing radiation emitted by the Earth's surface from escaping from the top of the atmosphere, thereby keeping the surface warmer than it would be by approximately 33°C. Thus, satellite observations of outgoing terrestrial radiation provide convincing evidence for the existence of a greenhouse effect of the atmosphere that intercepts a substantial fraction of the upward flux of terrestrial radiation emitted by the Earth's surface.

Figure 1.5 illustrates schematically the greenhouse effect of the atmosphere. In this figure, the slanted solid line indicates the idealized temperature profile of the troposphere, where temperature decreases almost linearly with height. The dot on the slanted line, located in the middle troposphere, indicates the average height of the layer of emission for the outgoing terrestrial radiation from the top of the atmosphere. As noted above, its temperature (T_A) is −18.7°C, which may be compared with +14.5°C—that is, the global mean temperature of the Earth's surface (T_s). As already noted, the difference between the two temperatures is 33°C, indicating that the atmosphere has a large greenhouse effect that warms the Earth's surface by this amount. In the following paragraphs, we shall attempt to explain why the atmosphere has this greenhouse effect and why it is so large.

Figure 1.6, which is a modified version of a figure presented by Peixoto and Oort (1992), illustrates how the atmosphere absorbs solar and terrestrial radiation. Figure 1.6a contains the normalized spectra of blackbody radiation at 255 K and 6000 K, which roughly mimic the spectra of outgoing terrestrial radiation and incoming solar radiation at the top of the atmosphere, respectively. As this figure shows, terrestrial radiation occurs at wavelengths that are mostly longer than 4 μm, whereas solar radiation involves wavelengths that are mostly shorter than 4 μm. Thus, it is reasonable to treat the transfer of terrestrial radiation in the atmosphere separately from that of solar radiation. Hereafter, we shall call the former "longwave radiation" to distinguish it from solar radiation with relatively short wavelengths.

Figure 1.6b and c illustrates the spectral distribution of absorption (%) by the cloud-free atmosphere. Although the clear atmosphere is almost transparent to the visible portion of solar spectrum at wavelengths between 0.3 and 0.7 μm, allowing the major fraction of incoming solar radiation to reach the Earth's surface, it absorbs very strongly over much of the spectral range of terrestrial longwave radiation, mainly owing to water vapor. In the so-called "atmospheric window" located between 7 and 20 μm, where water vapor is relatively transparent, carbon dioxide, ozone, methane, and nitrous oxide absorb very strongly around wavelengths of 15, 9.6, 7.7, and 7.8 μm, respectively, as shown in figure 1.6d. Although not shown in this figure, chlorofluorocarbons (CFCs) also absorb strongly in the 7–13 μm range, as pointed out, for example, by Ramanathan (1975). Although these greenhouse gases are minor constituents of the atmosphere, as shown in table 1.1, they collectively absorb and emit over much of the spectral range of terrestrial longwave radiation, exerting a powerful greenhouse effect as described below.

Radiative transfer from the Earth's surface and in the atmosphere obeys Kirchhoff's law, as described briefly in the box Blackbody Radiation and Kirchhoff's Law on page 8. This law requires that, at each wavelength, the absorptivity of a substance is equal to its emissivity, which is defined as the

Blackbody Radiation and Kirchhoff's Law

Let's consider a medium inside a perfectly insulated enclosure with a black wall that absorbs 100% of incident radiation. Assume that this system has reached the state of thermodynamic equilibrium characterized by uniform temperature and isotropic radiation. Because the wall is black, the radiation emitted by the system to the wall is absorbed completely. On the other hand, the outgoing emission from the wall is identical in magnitude to the incoming radiation. Radiation within the system is referred to as blackbody radiation, which depends only on temperature and wavelength according to the so-called Planck function. The spectral distributions of the normalized Planck function are shown in the right and left sides of figure 1.6a for 255 K and 6000 K, which approximate the equivalent emission temperature of the Earth and Sun, respectively. Summing up over all frequencies, the blackbody radiation depends only upon temperature and is proportional to the fourth power of the absolute temperature (in Kelvin) according to the Stefan-Boltzmann law of blackbody radiation.

In order to maintain the thermodynamic equilibrium in the enclosure, it is necessary for the wall and the medium to emit and absorb equal amounts of radiation, maintaining the radiative heat balance. This implies that, for a given wavelength, the emissivity, defined as the ratio of the emission to the Planck function, is equal to the absorptivity, defined as the ratio of the absorption to incident radiation. The equality between absorptivity and emissivity was first proposed by Kirchhoff in 1859.

Kirchhoff's law requires the condition of thermodynamic equilibrium, such that uniform temperature and isotropic radiation are achieved. Obviously, the radiation field of the Earth's atmosphere is not isotropic and its temperature is not uniform. However, in a localized volume below about 40 km, to a good approximation, it may be considered to be locally isotropic with a uniform temperature, in which energy transitions are determined by molecular collisions. It is in the context of this local thermodynamic equilibrium that Kirchhoff's law is applicable to the atmosphere. For further analysis of this subject, see Goody and Yung (1989).

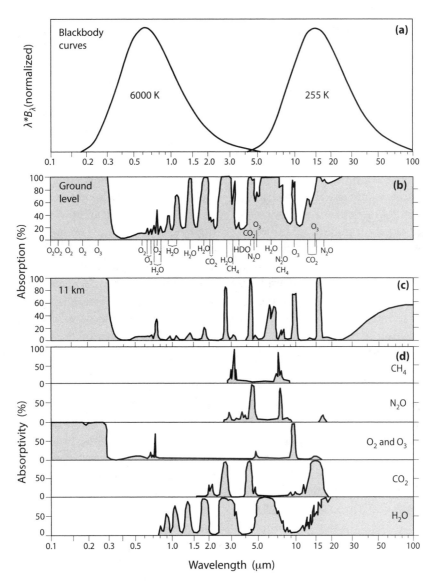

FIGURE 1.6 Spectra of blackbody emission and absorption by the atmosphere and its constituents. (*a*) Normalized spectra of blackbody radiation at 6000 K and 255 K; (*b*) absorption spectra for the entire vertical extent of the atmosphere and (*c*) for the portion of the atmosphere above 11 km; and (*d*) absorption spectra for the various atmospheric gases between the top of the atmosphere and the Earth's surface. CH$_4$, methane; CO$_2$, carbon dioxide; H$_2$O, water; HDO, hydrogen-deuterium oxide (i.e., heavy water); N$_2$O, nitrous oxide; O$_2$, oxygen; O$_3$, ozone; B$_\lambda$, blackbody emission at wavelength λ. From Peixoto and Oort (1992).

ratio of the actual emission to the theoretical emission from a blackbody. Because the Earth's surface behaves almost as a blackbody, it has an absorptivity that is close to one, absorbing almost completely the downward flux of longwave radiation that reaches it. In keeping with Kirchhoff's law, the Earth's surface emits an upward flux of longwave radiation almost as a blackbody would. As this upward flux penetrates into the atmosphere, it is depleted owing to the absorption by greenhouse gases, but it is also accreted because of the emission from these gases. The upward flux decreases or increases with height, depending upon whether the depletion is larger than the accretion, or vice versa.

For example, if the atmosphere were isothermal, **these** two opposing effects would exactly cancel, as they would do inside a blackbody enclosure with homogeneous and isotropic radiation, thus yielding an upward flux of radiation that would be constant with height. However, if temperature decreases with increasing height, as it does in the troposphere, the accretion of the upward flux due to emission is smaller than its depletion by absorption of the flux from below. Thus, the upward flux decreases with increasing height owing to the difference between emission and absorption in the atmosphere. In short, the atmosphere as a whole traps a substantial fraction of the upward flux of longwave radiation emitted by the Earth's surface before it reaches the top of the atmosphere. This trapping is often called the atmospheric greenhouse effect.

The atmospheric greenhouse effect is attributable not only to well-mixed greenhouse gases, but also to cloud cover. As will be described in chapter 6, clouds emit and absorb longwave radiation, acting almost as a blackbody if they are sufficiently thick. Clouds account for about 20% of the total greenhouse effect of the atmosphere. But the greenhouse effect of clouds is not the only effect on the Earth's radiation balance. Clouds also reflect incoming solar radiation because their albedo is higher than that of most underlying surfaces. When evaluated on a global and annual mean basis, the reflection of incoming solar radiation by clouds outweighs their greenhouse effect, and thus clouds exert a net cooling effect upon the heat balance of the planet (e.g., Hartmann, 2016; Ramanathan et al., 1989).

In summary, the atmosphere as a whole absorbs a major fraction of the upward flux of longwave radiation emitted by the Earth's surface. On the other hand, the atmosphere also emits longwave radiation, and the upward flux from the atmosphere partially compensates for the depletion of the upward flux due to absorption. Since Kirchhoff's law requires the absorptivity of the atmosphere to be identical to its emissivity at all wavelengths, the absorption of the upward flux emitted by the relatively warm surface is substantially larger than the emission of the upward flux by the relatively cold atmosphere. Thus, the atmosphere has a greenhouse effect that reduces substantially the upward flux of longwave radiation emitted by the Earth's surface before it reaches the top of the atmosphere, helping to maintain a warm and habitable planet.

Global Warming

So far, we have explained why the atmosphere has a so-called greenhouse effect that traps a substantial fraction of the upward flux of longwave radiation emitted by the Earth's surface. Here we attempt to explain why temperature increases at the surface and in the troposphere as the concentration of a greenhouse gas (e.g., CO_2) increases in the atmosphere.

As described in the preceding section, the effective emission temperature of the Earth is $-18.7°C$, which is much closer to the global mean temperature of the middle troposphere than that of the Earth's surface. The effective emission temperature is much lower than the Earth's surface temperature because a major fraction of longwave radiation emitted by the surface is absorbed before it reaches the top of the atmosphere. On the other hand, much of the upward flux of the longwave radiation emitted by the colder upper troposphere reaches the top of the atmosphere because the absorption by the overlying layers of the atmosphere is small. Thus, the effective emission level of outgoing longwave radiation is located in the middle troposphere, where temperature is much colder than the Earth's surface.

Using the terminology of quantum mechanics provides another way of visualizing the effect of greenhouse gases on longwave radiative transfer in the atmosphere. If we think of radiation taking the form of photons, the probability of a terrestrial photon escaping from the top of the atmosphere is reduced by the presence of greenhouse gases lying above the level from which that photon is emitted. Thus, photons emitted from the Earth's surface are much less likely to reach the top of the atmosphere than photons emitted from higher levels in the atmosphere. If we imagine each photon to be "tagged" with the temperature at which it was emitted, the distribution of photons reaching the top of the atmosphere will be centered at a temperature that is much lower than the surface temperature; that is, the effective emission temperature.

If the concentration of a greenhouse gas such as CO_2 increases in the atmosphere, the infrared opacity of the air increases, thereby enhancing the absorption of longwave radiation in the atmosphere. Thus the absorption of the upward flux of longwave radiation from the lower layer of the atmosphere is larger than that of the flux from the higher layer, owing mainly to the difference in the optical thickness of the overlying layer of the atmosphere. In other words, more photons from the surface and lower atmosphere will be prevented from reaching the top of the atmosphere. For this reason, the effective height of the layer from which the outgoing longwave radiation originates increases as the atmospheric greenhouse gas concentration increases. Because the effective emission level of the outgoing radiation is located in the troposphere, where temperature decreases with increasing height, the temperature of the effective emission level decreases as it moves upward, thereby reducing the outgoing longwave radiation from the top of the atmosphere.

The physical processes involved in the response of longwave radiation to increasing greenhouse gases are illustrated in figure 1.7. As in figure 1.5, the slanted solid line indicates schematically the vertical temperature profile in the troposphere, where temperature decreases almost linearly with height. Dot A on the slanted line indicates the effective emission level for outgoing longwave radiation from the top of the atmosphere. (In other words, half the photons reaching the top of the atmosphere are emitted from below this level and half from above this level.) As indicated by the arrow that originates from dot A, the emission level moves upward in response to the increase in concentration of an atmospheric greenhouse gas (e.g., CO_2), as explained above. Thus the temperature of the effective emission level decreases, reducing the outgoing longwave radiation from the top of the atmosphere.

A change in the concentrations of greenhouse gases (e.g., carbon dioxide and water vapor) affects not only the upward flux of the outgoing longwave radiation from the top of the atmosphere but also the downward flux that reaches the Earth's surface. If the concentration of atmospheric greenhouse gases increases, the increase in the infrared opacity of the air enhances the absorption of longwave radiation in the atmosphere. Thus, the absorption of the downward flux from the higher layer of the atmosphere increases more than the absorption of the flux from the lower layer. Consequently, there is a downward shift of the effective level of emission from which the downward flux originates as the atmospheric greenhouse-gas concentration increases. Because temperature increases with decreasing height in the troposphere, as indicated by the slanted line in figure 1.7, the temperature at the effective emission level of the downward flux, signified by dot B, also increases as it moves downward, thereby increasing the downward flux of longwave radiation that reaches the Earth's surface.

The radiative response of the surface-troposphere system to an increase in greenhouse gases can be regarded as the net result of two related

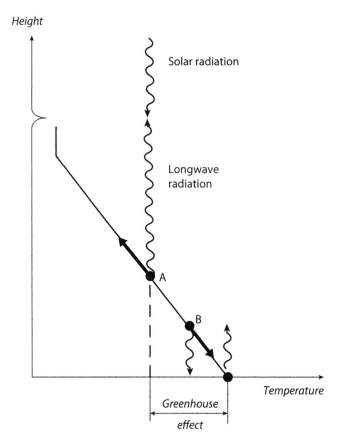

FIGURE 1.7 Schematic diagram illustrating the upward shift of the average height of the layer of emission for the top-of-the-atmosphere flux of longwave radiation (indicated by dot *A*), in response to the increase in concentration of a greenhouse gas in the atmosphere. The diagram also illustrates the downward shift of the average height of the layer of emission for the downward flux the longwave radiation at the Earth's surface (indicated by dot *B*), in response to the increase in concentration of a greenhouse gas. See the caption of figure 1.5 for further details.

processes. The first process involves an increase in the downward flux of longwave radiation that increases the temperature of the Earth's surface. Over a sufficiently long period of time, the surface returns to the overlying troposphere practically all the radiative energy it receives, with thermal energy being transferred upward through moist and dry convection, longwave radiation, and the large-scale circulation. Thus temperature increases in the troposphere, as well as at the Earth's surface. If this warming were to occur in the absence of any other changes, it would result in an increase in the outgoing longwave radiation from the top of the atmosphere.

The second process involves the upward flux of longwave radiation at the top of the atmosphere in response to an increase in atmospheric greenhouse gas concentration. If the amount of greenhouse gases were to increase without allowing the temperature of the surface-troposphere system to change, the upward flux of longwave radiation at the top of the atmosphere would decrease, as explained earlier. To maintain the radiative heat balance of the planet as a whole, the surface-troposphere system warms just enough for the effects of these two processes to balance, such that the top-of-atmosphere flux of outgoing longwave radiation remains unchanged despite the warming.

An important factor that affects the magnitude of global warming is the positive feedback process that involves water vapor. As noted already, water vapor is the most powerful greenhouse gas in the atmosphere. It absorbs and emits strongly over much of the spectral range of terrestrial longwave radiation (figure 1.6d) and is mainly responsible for the powerful greenhouse effect of the atmosphere. In contrast to long-lived greenhouse gases such as carbon dioxide, water vapor has a short residence time of a few weeks in the atmosphere, where it is added rapidly through evaporation from the Earth's surface (e.g., from the ocean) and is depleted through condensation and precipitation. Thus, the absolute humidity of air is bounded by saturation, preventing large-scale relative humidity from exceeding 100%. Since the saturation vapor pressure of air increases with increasing temperature according to the Clausius-Clapeyron equation, the absolute humidity of air usually increases with increasing temperature, thereby enhancing the atmospheric greenhouse effect. The positive feedback effect between temperature and the greenhouse effect of the atmosphere is called "water vapor feedback," and will be discussed further in chapter 6. Water vapor feedback magnifies the global warming that is induced by the increase in long-lived greenhouse gases such as carbon dioxide, methane, nitrous oxide, and CFCs.

Global warming involves other responses of the climate system in addition to the changes in temperature. Because the saturation vapor pressure of air in contact with a wet surface (e.g., ocean) increases at an accelerated pace with increasing surface temperature, according to the Clausius-Clapeyron equation, the warming of the Earth's surface enhances evaporation from the surface to the overlying troposphere, where relative humidity hardly changes, as will be discussed in subsequent chapters. The global-scale increase in the rate of evaporation in turn increases the rate of precipitation, thereby satisfying the water balance of the atmosphere in the long run. This is the main reason why the global mean rates of evaporation and precipitation increase by equal magnitudes as global warming proceeds, accelerating the pace of the global water cycle, as will be discussed further in chapter 10.

In this chapter, we have explained the processes responsible for the atmospheric greenhouse effect, which is essential for maintaining a warm and habitable climate at the Earth's surface. We have also explained why temperature increases at the Earth's surface and the global water cycle accelerates as the concentration of CO_2 increases in the atmosphere. In the remainder of this book, we shall introduce various studies exploring the physical mechanisms that control changes in climate, not only during the industrial present, but also during the geologic past. We will introduce these studies in roughly historical order; we begin by introducing the early pioneering studies in the following chapter.

CHAPTER 2

Early Studies

In this chapter we introduce the early studies of the greenhouse effect of the atmosphere and global climate change, conducted during the nineteenth and early twentieth centuries. We begin with the pioneering studies of Fourier, Tyndall, Arrhenius, and Hulbert.

The Heat-Trapping Envelope

The existence of the greenhouse effect of the atmosphere described in the preceding chapter was conjectured, perhaps for the first time, by the well-known mathematical physicist Jean-Baptiste Fourier. In essays published in 1824 and 1827 (Fourier, 1827; Pierrehumbert, 2004a, b), Fourier referred to an experiment conducted by Swiss scientist Horace-Bénédict de Saussure. In this experiment, de Saussure lined a container with blackened cork and inserted into the cork several panes of transparent glass, separated by intervals of air. Midday sunlight could enter at the top of the container through the glass panes. The temperature became more elevated in the most interior compartment of this device. Fourier speculated that the atmosphere could form a stable barrier like the glass panes, trapping a substantial fraction of the upward flux of terrestrial radiation emitted by the Earth's surface before it reached the top of the atmosphere, despite being almost transparent to incoming solar radiation as indicated in figure 1.6b. Although Fourier did not elaborate on the specific mechanism involved in the trapping described in chapter 1, it is quite remarkable that he correctly conjectured the existence of the greenhouse effect of the atmosphere based upon the simple experiments conducted by de Saussure. (For interesting commentaries on the papers by Fourier and de Saussure, see chapter 1 of *The Warming Papers*, edited by Archer and Pierrehumbert [2011].)

Fourier, however, did not identify the constituents of the atmosphere that allow it to act as a heat-trapping envelope. Irish physicist John Tyndall successfully identified these gases and evaluated the relative magnitudes of their contributions to the greenhouse effect of the atmosphere. His measuring device, which used thermopile technology, is an early landmark in the history of absorption spectroscopy of gases. Tyndall (1859, 1861) concluded that, although major constituents of the air such as nitrogen and oxygen are transparent to longwave radiation, minor constituents such as water vapor, carbon dioxide, methane, nitrous oxide, and ozone absorb and emit longwave radiation (figure 1.6d), exerting the greenhouse effect. He found that, among these absorbing gases, water vapor is the strongest absorber in the atmosphere, followed by carbon dioxide. They are the principal gases controlling surface air temperature. The relative contributions of these gases to global warming have been the subject of quantitative evaluation by Wang et al. (1976) and Ramanathan et al. (1985).

The First Quantitative Estimate

In 1894, Svante Arrhenius of Sweden (figure 2.1) gave a very provocative lecture at a meeting of the Stockholm Physical Society, proposing that a change in the atmospheric CO_2 concentration by two to three times may be enough to induce a change in climate comparable in magnitude to the glacial-interglacial difference in global mean surface temperature. In a paper published the following year, Arrhenius (1896) described in specific detail how he conducted his study, which turned out to be relevant not

FIGURE 2.1 Svante Arrhenius (1859–1927).

only to the large glacial-interglacial transition of climate (chapter 7), but also the global warming that is currently taking place. Here, we describe his truly pioneering study, which used a simple climate model to estimate the magnitude of the surface temperature change resulting from a change in atmospheric CO_2 concentration.

For this purpose, Arrhenius formulated two equations that represent the heat balance of the atmosphere and that of the Earth's surface at each latitude and during each season of the year. In his formulation, the heat balance of the atmosphere is maintained among the following components:

- cooling and heating due to the emission and absorption of longwave radiation, respectively
- heating due to the absorption of solar radiation
- heating due to the net upward flux of heat from the Earth's surface to the atmosphere
- heating or cooling due to meridional heat transport by large-scale circulation in the atmosphere

The heat balance of the Earth's surface is maintained among the following components, implicitly assuming that the surface has no heat capacity and returns all of the radiative heat energy it receives in sensible and latent heat:

- cooling and heating due to the emission and absorption of longwave radiation, respectively
- heating due to the absorption of solar radiation
- cooling due to the net upward flux of heat from the Earth's surface to the atmosphere

From the equation of the heat balance of the atmosphere and that of the Earth's surface described above, Arrhenius obtained a formula that relates the temperature of the Earth's surface to the atmospheric concentration of CO_2. Using the formula thus obtained, he computed the change in surface temperature at various latitudes and during each season in response to a change in the CO_2 concentration of the atmosphere. The positive feedback effect of water vapor described in chapter 1 was incorporated iteratively, keeping relative humidity unchanged in the atmosphere as temperature changed in each successive iteration. In addition to water vapor feedback, he incorporated the positive feedback effect of snow cover that retreats poleward as temperature increases, thereby enhancing the warming at the Earth's surface. It is quite impressive that Arrhenius identified two of the most important positive feedback effects and incorporated them into his computation. It should be noted here, however, that Arrhenius assumed that the horizontal and vertical heat flux do not change despite the change in temperature at the Earth's surface and in the atmosphere, greatly simplifying his computation. This assumption implies that the magnitude of the surface temperature change he obtained is controlled solely by radiative processes and does not depend on heat transport by the large-scale circulation and vertical convective heat transfer in the atmosphere.

Averaging globally and annually the surface temperature change thus obtained, Arrhenius found that the global mean surface temperature would increase by 5°C–6°C in response to a doubling of the atmospheric CO_2 concentration. The magnitude of the global warming thus obtained is quite large and lies at the upper end of the sensitivity range of the current climate models (Flato et al., 2013). As discussed below, the large sensitivity of his model is attributable mainly to the unrealistically large absorptivity/emissivity of CO_2 that Arrhenius used in his computation.

In order to estimate the absorption spectra of water vapor and CO_2, Arrhenius used the record of radiation from the Moon obtained by astronomer and physicist Samuel Langley (1889). As pointed out in a detailed analysis by Ramanathan and Vogelmann (1997), the CO_2 absorptivity Arrhenius used is too large by a factor of ~2.5. They suggested that this discrepancy is attributable mainly to the error involved in guessing the magnitude of the CO_2 absorptivity from Langley's observations over the spectral range, where the absorption bands of CO_2 overlap with those of water vapor. When they repeated Arrhenius's computation in the absence of albedo feedback using modern CO_2 absorptivity data, they found that surface temperature increases by only ~2°C in response to the CO_2 doubling. Although this is much smaller than the 5°C–6°C that Arrhenius obtained, it is similar to the sensitivity of one-dimensional models of the surface-atmosphere system that use modern CO_2 absorptivity data.

As discussed already, the magnitude of the atmospheric greenhouse effect depends not only on the vertical distribution of greenhouse gases but also on the vertical temperature structure, which depends on both convective and radiative heat transfer. To obtain a reliable estimate of global warming, it is therefore desirable to use a multilayer radiative-convective model of the atmosphere rather than the one-layer model that Arrhenius used. The initial attempt to use such a model was made by E. O. Hulbert (1931). He developed a vertical-column model of the atmosphere, which consisted of the troposphere in radiative-convective equilibrium and the stratosphere in radiative equilibrium. He was very encouraged to find that his model simulated successfully the global mean temperature of the Earth's surface. Using this model, Hulbert estimated the magnitude of surface temperature change that results from a given change in atmospheric CO_2 concentration. He found that the temperature of the Earth's surface increases by 4°C in response to a doubling of the atmospheric CO_2 concentration in the absence of positive water vapor feedback. Realizing that the magnitude of the warming would have been even larger if this feedback had been taken into consideration, he concluded, in agreement with Arrhenius, that the CO_2 theory of the ice ages was credible, despite various objections that had been raised against it.

The warming of 4°C (in response to CO_2 doubling) obtained by Hulbert happens to be similar to the 3.8°C warming that Ramanathan and Vogelmann (1997) obtained, using a version of the Arrhenius's model, in the absence of the water vapor and snow albedo feedbacks. Both of these values are about three times as large as the ~1.2°C that one gets from a radiative-convective model without these feedbacks, using modern CO_2 absorptivity. It is therefore likely that Hulbert used CO_2 absorptivity data that were too large by a factor of almost three, as Arrhenius did. Had Hulbert incorporated water vapor feedback into his computation, he would have found a surface temperature increase of as much as 6°C in response to a doubling of the atmospheric CO_2 concentration.

Hulbert's study is a natural extension of the study of Arrhenius. It uses for the first time a radiative-convective model of the atmosphere, which resolves the vertical temperature structure, to estimate the magnitude of global warming. Unfortunately, Hulbert's study was overlooked or disregarded for a long time in favor of a very simple approach proposed by British engineer Guy Stewart Callendar, which we will discuss in the next section. Although Hulbert's study has a serious flaw as discussed above, it is a truly groundbreaking contribution that predates by more than three decades the radiative-convective model study of Manabe and Wetherald (1967) that will be described in chapter 3.

A Simple Alternative

Several decades after the publication of Arrhenius's study, Callendar made a renewed attempt to estimate the change of surface temperature that results from a change in atmospheric CO_2 concentration. The main motive of his study, however, was different from that of Arrhenius's. Arrhenius was mainly interested in exploring the role of greenhouse gases in producing the glacial-interglacial contrast in climate. But Callendar began his paper with the following assertion: "Few of those familiar with the natural heat exchanges of the atmosphere, which go into the making of our climate and weather, would be prepared to admit that the activities of man could have any influence upon phenomena of so vast a scale. In the following paper, I hope to show that such influence is not only possible but is occurring at the present time" (Callendar, 1938, 223).

Although Callendar was aware of Arrhenius's study described above, he attempted to improve the estimate of the CO_2-induced change in surface temperature using absorptivities of CO_2 and water vapor that are more realistic than those used by Arrhenius. Callendar, however, employed a very simple method that is based solely upon the radiative heat balance of the Earth's surface, which we describe in the remainder of this chapter.

As noted in chapter 1, a change in atmospheric greenhouse gas concentration results in a change in the intensity of the downward flux of longwave radiation at the Earth's surface. If the atmospheric greenhouse gas concentration increases, for example, the infrared opacity of air also increases, thereby enhancing the absorption of longwave radiation in the atmosphere. This implies that the absorption of the downward flux of longwave radiation from higher layers of the atmosphere increases more than that of the flux from lower layers. Thus, the effective height of the layer from which the downward flux originates decreases. Since temperature increases with decreasing height in the troposphere, where the downward flux of longwave radiation originates, this implies that the downward flux of longwave radiation increases at the Earth's surface as the concentration of CO_2 increases in the atmosphere. To maintain the heat balance of the Earth's surface, it is therefore necessary that the increase in downward flux is compensated by an equal increase in the upward flux of longwave radiation, if other things remain unchanged. In his study, Callendar estimated the magnitude of the surface temperature change needed to keep the net upward flux of longwave radiation unchanged at the Earth's surface despite an increase in downward flux due to the increase in atmospheric CO_2 concentration.

The net upward flux of longwave radiation (E) at the Earth's surface is defined here as the difference between the upward flux (U) and downward flux (D) as expressed by:

$$E = U - D, \tag{2.1}$$

where E is usually positive and is expected to increase with increasing surface temperature.

The perturbation equation of the net upward flux of longwave radiation at the Earth's surface may be expressed by:

$$dE(C, T_s) = (\partial E/\partial C) \cdot dC + (\partial E/\partial T_s) \cdot dT_s, \qquad (2.2)$$

where C is the atmospheric CO_2 concentration, and T_s is the global mean temperature of the Earth's surface. Assuming that the surface heat balance is maintained, $dE(C, T_s) = 0$, which means that $(\partial E/\partial C) = -(\partial D/\partial C)$. One can then derive the relationship between the changes in CO_2 concentration and surface temperature, as expressed by:

$$dT_s = [(\partial D/\partial C)/(\partial E/\partial T_s)] \cdot dC. \qquad (2.3)$$

Using equation (2.3) thus obtained, Callendar estimated the magnitude of surface temperature change that results from a given change in atmospheric CO_2 concentration. He assumed that the vertical gradient of temperature is constant and does not depend upon surface temperature. This method led to the finding that surface temperature increases by about 2°C in response to the doubling of atmospheric CO_2. Although it is less than half of the 5°C–6°C that Arrhenius obtained earlier, Callendar's result appeared to be consistent with the observed increase of the global mean surface temperature over several decades in the late nineteenth and early twentieth centuries.

A few decades after the publication of Callendar's study, several authors attempted to repeat the study, incorporating various factors that had been neglected by Callendar (Kaplan, 1960; Kondratiev and Niilisk, 1960; Möller, 1963; Plass, 1956). For example, Plass found that surface temperature increases by 3.6°C in response to a doubling of the CO_2 concentration. Kaplan obtained ~1.5°C, taking into consideration the effect of clouds on the change in the downward flux of longwave radiation. Using the best CO_2 absorptivity available at that time (e.g., Yamamoto and Sasamori, 1961), Möller (1963) obtained 1°C, which is about half as large as the 2°C originally obtained by Callendar. Realizing that the results presented above were obtained without accounting for water vapor feedback, Möller repeated the computation incorporating the feedback and obtained a quite surprising result.

As explained in chapter 1, a warming of the Earth's surface induced by an increase in greenhouse gases is accompanied by an increase in temperature throughout the troposphere and an increase in absolute humidity, keeping relative humidity essentially unchanged. The increase in absolute humidity leads to an additional increase in the infrared opacity of the troposphere, reducing further the average height of the layer from which the downward flux of longwave radiation originates. Because temperature increases with decreasing height in the troposphere as noted above, the downward flux of longwave radiation increases, thereby enhancing the warming at the Earth's surface through the positive feedback effect of water vapor. To incorporate this effect, Möller modified equation (2.3), replacing $\partial E/\partial T_S$ by dE/dT_S, and obtained the following equation, which describes the relationship between dT_S and dC in the presence of the water vapor feedback:

$$dT_S = [(\partial D/\partial C)/(dE/dT_S)] \cdot dC, \qquad (2.4)$$

where dE/dT_S depends not only on T_S but also on W (i.e., the total moisture content of the atmosphere). Given that $(\partial E/\partial W) = -(\partial D/\partial W)$, it may be expressed by:

$$dE/dT_S = \partial E/\partial T_S - \partial D/\partial W \cdot (dW/dT_S). \qquad (2.5)$$

Since the downward flux of longwave radiation increases with increasing W, which increases with increasing surface temperature as explained above, the second term, $\partial D/\partial W \cdot (dW/dT_S)$, on the right side of equation (2.5) is positive and compensates for the first term, which is positive as noted already, thereby reducing the strength of the radiative feedback (i.e., dE/dT_S) that operates on the perturbation of the global mean surface temperature. Assuming that surface temperature is 15°C and relative humidity of air is fixed at 77%, for example, the two terms on the right side of equation (2.5) compensate for each other almost completely, yielding a very small negative value of dE/dT_S. In other words, the net upward flux of longwave radiation decreases slightly with increasing surface temperature. This implies that, in the presence of water vapor feedback, the net upward flux of longwave radiation cannot compensate for the increase in the downward flux due to the increase in atmospheric CO_2 concentration. Inserting the value of dE/dT_S thus obtained into equation (2.4), one gets a large cooling of as much as 6°C in response to a doubling of atmospheric CO_2. This irrational result is obtained because longwave feedback is positive and equation (2.4) is no longer valid.

One of the important components of heat balance at the Earth's surface is the upward flux of sensible heat and latent heat of evaporation to the atmosphere. When the downward flux of longwave radiation increases at the Earth's surface owing to an increase in atmospheric CO_2 concentration, the temperature increases at the Earth's surface, thereby increasing the upward sensible and latent heat flux to the atmosphere. To estimate the magnitude of temperature change at the Earth's surface, Callendar assumed, however, that the upward flux of sensible and latent heat does not change, despite the change in surface temperature. Thus, he neglected one of the important processes that contribute to the heat balance of the Earth's surface. This is the main reason why Möller, who used a surface heat balance approach similar to Callendar's, encountered the difficulty described above. Although Arrhenius also assumed incorrectly that the upward sensible and latent heat flux does not change, he did not experience a similar difficulty because the vertical temperature gradient in the atmosphere is permitted to increase in his model. The strengthened vertical gradient results in an increase in the net upward flux of longwave radiation at the Earth's surface, which helps to compensate for the increase in the downward flux that results from the increase in CO_2 and water vapor. On the other hand, the vertical temperature gradient is held fixed in Callendar's model, making it impossible for the net upward flux of longwave radiation to increase enough to compensate for the increase in downward flux that results from the increase in both of these gases in the atmosphere.

We note here that Newell and Dopplick (1979) made a renewed attempt to estimate the change in surface temperature in response to a doubling of atmospheric CO_2 concentration, using a heat balance model of the Earth's surface that incorporates the effect of heat exchange between the surface and the atmosphere. They found that the magnitude of the increase in surface temperature is less than 0.25°C. As discussed by Watts (1981), for example, the small response of their model is attributable in no small part to the unrealistic assumption that the temperature and absolute humidity of air in the near surface layer of the atmosphere do not change in response to the increase in CO_2 concentration, thereby severely constraining the magnitude of temperature change at the Earth's surface.

To obtain a reliable estimate of global warming, it is desirable to construct a model in which the heat exchange between the Earth's surface and the atmosphere is computed without making an artificial assumption as Callendar did. An approach that did not require this assumption involved the adaptation of a radiative-convective model (such as that developed by Hulbert) for use in the study of global warming. In chapter 3 we shall present such a study conducted in the early 1960s, when accurate measurements of the absorptivity of greenhouse gases became available and simple schemes for reliably estimating radiative heat transfer (e.g., Goody, 1964; Yamamoto, 1952) had been developed.

One-Dimensional Model

Radiative-Convective Equilibrium

It was the early 1960s when a one-dimensional (1-D) vertical column model of the atmosphere was developed at the Geophysical Fluid Dynamical Laboratory (GFDL) of what was then the United States Weather Bureau (now the National Oceanic and Atmospheric Administration). This 1-D model of radiative-convective equilibrium was developed as a first step toward the development of the three-dimensional (3-D) general circulation model of the atmosphere that will be described in the following chapter. Nonetheless, the 1-D model has been very useful for exploring the roles of various greenhouse gases (e.g., water vapor, carbon dioxide, and ozone) in maintaining the thermal structure of the atmosphere. Here we describe the structure of the model as constructed by Manabe and Strickler (1964) and evaluate its performance in simulating the vertical distribution of temperature in the atmosphere. We then describe, in the latter half of this chapter, how it is used for estimating the response of temperature in the atmosphere and at the Earth's surface to changes in the atmospheric concentration of CO_2.

In the radiative-convective model, four processes are in operation. They are solar radiation, longwave radiation, atmospheric convection, and heat exchange between the Earth's surface and the atmosphere. Given the vertical distribution of clouds and the concentrations of greenhouse gases such as water vapor, carbon dioxide, and ozone in the atmosphere, the model is initialized with a prescribed vertical temperature profile and computes numerically the rate of temperature change due to (1) absorption of solar radiation, (2) emission and absorption of longwave radiation, (3) upward heat flux from the Earth's surface to the overlying troposphere, and (4) upward convective heat transfer in the troposphere.

The heat balance at the Earth's surface is maintained among solar radiation, longwave radiation, and the heat flux from the Earth's surface to the atmosphere, as noted above. Although the surface heat flux consists of the fluxes of sensible and latent heat, these two fluxes are not distinguished from each other in the model but are treated together as a single heat flux. In the globally averaged model described here, this simplification may be justified because practically all the water vapor evaporated from the Earth's surface eventually condenses, transforming into sensible heat as it is transported upward in the troposphere. It should be noted here, however, that there are other components of the surface heat balance, such as vertical mixing of heat in the ocean and heat conduction in the soil. Although these fluxes could be large over a short period of time, they are generally quite small when averaged over a sufficiently long period of time. Thus, it is assumed here that the Earth's surface returns to the overlying atmosphere all of the energy it receives.

In the model, convection is represented by a procedure called "convective adjustment." When the vertical lapse rate of temperature becomes supercritical, this procedure adjusts it to the convectively neutral rate, keeping the total potential energy (i.e., the sum of potential and internal energy) unchanged. The critical lapse rate of convection was chosen here to be 6.5°C km^{-1}, which is close to the observed global mean lapse rate of temperature in the troposphere. This lapse rate is not very different from the moist adiabatic temperature profile that follows an ascent of a saturated air parcel in the atmosphere, underscoring the predominant influence of deep moist convection in low and middle latitudes that will be discussed in chapter 4.

In the version of the model constructed by Manabe and Strickler, the atmosphere is subdivided into 18 layers of unequal thickness and a temperature is determined for each layer. The numerical time integration of the model is performed in small time steps, starting from an isothermal temperature profile. At each time step, the model computes the temperature change due to both solar and longwave radiation. In the lowest layer, the temperature change also includes the contribution of the heat flux from the Earth's surface. The temperature changes thus obtained are added to the temperature at the end of preceding time step.

The vertical temperature profile thus obtained is modified further through the convective adjustment procedure, which adjusts the vertical lapse rate of temperature to the convectively neutral rate whenever it becomes supercritical, while conserving the total energy of the vertical air column. Upon completion of the convective adjustment procedure, the time integration of the model proceeds to the next time step, repeating the process until the state of the model atmosphere ceases to change and the net incoming solar radiation at the top of the atmosphere becomes practically identical to the outgoing longwave radiation.

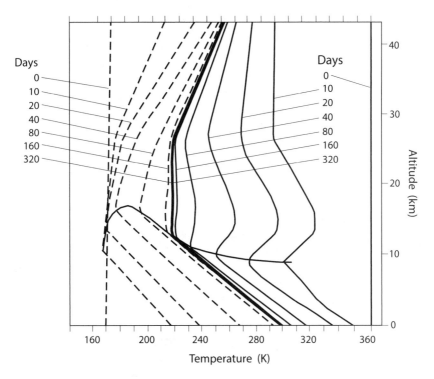

FIGURE 3.1 The vertical profile of temperature as it approaches radiative-convective equilibrium from cold and warm isothermal initial conditions. The dashed and solid lines show the approaches from cold and warm isothermal atmospheres with temperatures of 170 K and 360 K, respectively; the thick solid line indicates the final profile. The thin quasi-horizontal solid line between 9 and 16 km altitude denotes the position of the tropopause. From Manabe and Strickler (1964).

Using this model, Manabe and Strickler (1964) obtained two states of the radiative-convective equilibrium for idealized, cloud-free atmosphere. Initializing the model with very warm and very cold isothermal atmospheric temperature profiles, they performed two numerical time integrations of the model. The globally averaged annual mean flux of incoming solar radiation was imposed at the top of the atmosphere throughout the course of the two integrations. The mean concentration of carbon dioxide and the vertical distributions of water vapor and ozone were prescribed based on observations and held fixed over time.

Figure 3.1 shows how the vertical profiles of the global mean temperature evolve in the atmosphere in these two time integrations. Despite the large difference between the two initial conditions, the simulated profiles become practically identical toward the end of the two 320-day time integrations. The final profile, illustrated by the thick solid line, consists of

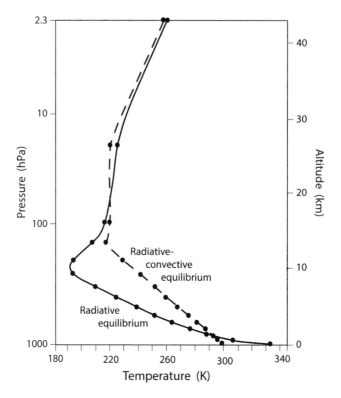

FIGURE 3.2 Vertical temperature profiles of cloud-free atmosphere in radiative equilibrium and in radiative-convective equilibrium. From Manabe and Strickler (1964).

a convective troposphere with constant vertical temperature gradient, an almost isothermal lower stratosphere, and an upper stratosphere where temperature increases gradually with height.

To evaluate the influence of convection upon the thermal structure of the atmosphere, the vertical profile of temperature in radiative-convective equilibrium is compared with the profile of radiative equilibrium (figure 3.2), in which the heat balance is maintained between solar radiation and longwave radiation in the absence of convection. The difference between the two profiles represents the contribution of convection to the temperature in the atmosphere and at the Earth's surface.

In the state of radiative equilibrium represented by the solid line in figure 3.2, the temperature is very high at the Earth's surface (333 K, or 60°C) but decreases sharply with height at a rate that is much larger than 6.5°C km^{-1}; that is, the critical lapse rate of temperature for convection. Because the cloud-free atmosphere is almost transparent to visible radiation with wavelengths between 0.4 and 0.7 μm, as shown in figure 1.6b, it allows a major fraction of incoming solar radiation to reach the Earth's surface, where it is absorbed or reflected. On the other hand, the atmosphere traps a major fraction of the upward flux of longwave radiation emitted by the Earth's surface, as explained in chapter 1. To maintain the balance between the incoming solar radiation and net outgoing longwave radiation without convection in the troposphere, it is necessary to maintain a very high temperature at the Earth's surface, as shown in figure 3.2.

In the state of radiative-convective equilibrium, on the other hand, the temperature is 300 K (27°C) at the Earth's surface and decreases linearly with height at the critical lapse rate for convective adjustment (i.e., 6.5°C km^{-1}). The surface temperature in this case is much lower than the temperature in radiative equilibrium, whereas the reverse is the case in the middle and upper troposphere. In short, convection transfers heat upward and is responsible for the formation of the troposphere beneath the stable stratosphere where convection is absent.

Figure 3.3a shows how the heat balance is maintained in the cloud-free atmosphere that satisfies the condition of radiative-convective equilibrium. In the troposphere, heat balance is maintained between heating due to convection and absorption of solar radiation and cooling due to longwave radiation. As described in chapter 1, water vapor and carbon dioxide emit and absorb longwave radiation over a wide range of the terrestrial longwave spectrum and are responsible for the net longwave cooling of the troposphere shown in figure 3.3a. On the other hand, water vapor has strong absorption bands of solar radiation between 0.8 and 4 μm (figure 1.6d) and is responsible for the solar heating of the troposphere shown in both panels of figure 3.3.

In the stratosphere, where convective heating is absent, the heat balance is maintained between the heating due to absorption of solar radiation and the net cooling due to longwave radiation, as shown in figure 3.3a. The solar heating is attributable mainly to ozone, which absorbs ultraviolet radiation very intensely at the short end of solar spectrum, at wavelengths shorter than 0.3 μm (figure 1.6d). On the other hand, carbon dioxide emits and absorbs longwave radiation very intensely around the wavelength of 15 μm (figure 1.6d) and is mainly responsible for the net longwave cooling of the stratosphere shown in figure 3.3a. Although water vapor also contributes to the cooling, its magnitude is smaller than that of carbon dioxide, as shown in figure 3.3b. In summary, the heat balance of the stratosphere is maintained essentially between the heating due to absorption of solar radiation by ozone and net cooling due to the emission and absorption of longwave radiation by carbon dioxide.

The state of radiative-convective equilibrium discussed above was obtained for the cloud-free atmosphere. In the actual atmosphere, however,

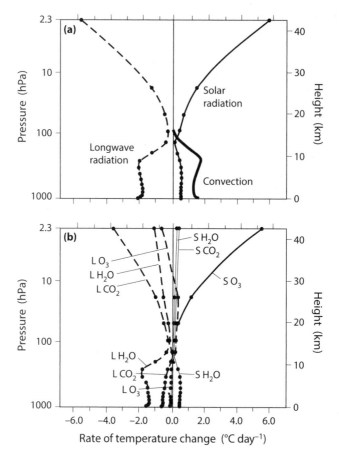

FIGURE 3.3 Vertical distributions of the components of heat budget of the cloud-free atmosphere in radiative-convective equilibrium. (*a*) The vertical distributions of the rates of temperature change (°C) due to convection, solar radiation, and longwave radiation. (*b*) Solid lines indicate the vertical profiles of the rate of temperature change (°C) due to the absorption of solar radiation by water vapor (S H_2O), carbon dioxide (S CO_2), and ozone (S O_3); dashed lines indicate the rates of temperature change due to the emission and absorption of longwave radiation by water vapor (L H_2O), carbon dioxide (L CO_2), and ozone (L O_3), respectively. From Manabe and Strickler (1964).

clouds cover about half of the globe. As noted in chapter 1, clouds exert a greenhouse effect that warms the Earth's surface through the absorption of longwave radiation. On the other hand, clouds reflect incoming solar radiation, exerting a cooling effect. Because the latter is larger than the former, clouds have a net cooling effect upon the heat budget of the planet. In order to simulate realistically the temperature of the Earth's surface, Manabe and Strickler repeated the computation, prescribing cloudiness based upon observed data compiled by London (1957). The equilibrium temperature thus obtained was 287 K (14°C) at the Earth's surface and

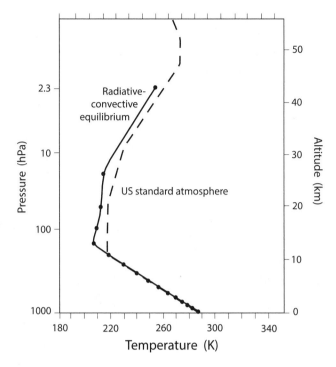

FIGURE 3.4 The vertical distribution of temperature in radiative-convective equilibrium obtained for the atmosphere with average cloudiness in middle latitudes. The vertical temperature distribution of the US standard atmosphere is also shown for comparison. From Manabe and Strickler (1964).

is similar to the observed global mean temperature. It is, however, about 13°C colder than the 300 K (27°C) temperature obtained for the cloud-free case, indicating the net cooling effect of clouds upon the global mean temperature at the Earth's surface.

In figure 3.4, the vertical distribution of temperature in radiative-convective equilibrium thus obtained is compared with the US standard atmosphere, which represents schematically the profile of the annual mean temperature over the mainland of the United States in middle latitudes. The model reproduces the temperature of the closely coupled surface-troposphere system quite well, although it underestimates the temperature of the lower stratosphere by several degrees.

The Response to CO_2 Change

Encouraged by the success of the 1-D radiative-convective model in simulating the vertical temperature profile of the atmosphere, Manabe and Wetherald (1967) used it to estimate the temperature change in response to a change in CO_2 concentration in the atmosphere. They made three simulations, each with a different CO_2 concentration. The first simulation, representing the control case, obtained the state of radiative-convective equilibrium for a CO_2 concentration of 300 ppmv, which was only a little smaller than the observed concentration at the time this study was conducted. Additional simulations were made with CO_2 concentrations of 600 and 150 ppmv, twice and half as large as the concentration in the control simulation, respectively. From the differences among the three states obtained in these simulations, they estimated the equilibrium response of temperature to the doubling and halving of the CO_2 concentration in the atmosphere.

For each concentration of atmospheric CO_2, the numerical time integration of the model was performed over the period of a few hundred days. To incorporate the positive feedback effect of water vapor, the absolute humidity of the atmosphere was adjusted continuously so that the distribution of relative humidity was held constant in the troposphere throughout the course of all three integrations. In the stratosphere, where convection is absent, the absolute humidity was held fixed at very small values, based on balloon observations conducted by Mastenbrook (1963).

Figure 3.5 illustrates the three vertical profiles of the temperature in radiative-convective equilibrium thus obtained. For the standard CO_2 concentration of 300 ppmv, surface temperature is 288.4 K (15°C), which is similar to the observed global mean surface temperature. In response to the doubling of the atmospheric CO_2 concentration to 600 ppmv, the temperature increases by 2.4°C not only at the Earth's surface but also in the entire troposphere, as discussed in chapter 1. On the other hand, temperature decreases by a similar magnitude of 2.3°C in response to the CO_2 halving from 300 to 150 ppmv.

In the CO_2-doubling experiment described above, CO_2 concentration increases by 300 ppmv, but in the CO_2-halving experiment, it decreases by only 150 ppmv. Although the magnitude of the change in CO_2 concentration is twice as large in the former than the latter, the magnitude of the temperature change is practically identical between the two experiments. The physics of radiative transfer is responsible for this nonlinear result. Because CO_2 absorptivity (or emissivity) is approximately proportional to the logarithm of the amount of CO_2, it is expected that the greenhouse effect of the atmosphere also changes in proportion to the logarithm of the atmospheric CO_2 concentration. Thus, the magnitude of the warming due to CO_2 doubling is similar to that of the cooling due to CO_2 halving, even though the magnitude of the change in CO_2 concentration is only half as large in the latter case.

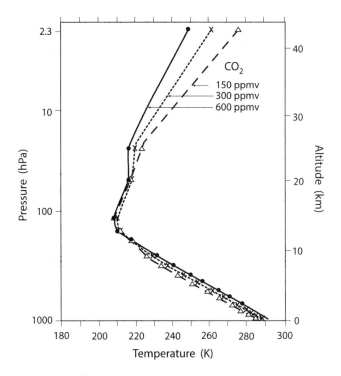

FIGURE 3.5 Vertical profiles of temperature in radiative-convective equilibrium obtained for the three different atmospheric concentrations of carbon dioxide: 150, 300, and 600 ppmv (parts per million by volume). From Manabe and Wetherald (1967).

Although the mechanism of temperature change in the surface-troposphere system was discussed in chapter 1, we shall repeat it briefly here for the convenience of the reader. In response to an increase in the atmospheric CO_2 concentration, for example, the downward flux of long-wave radiation increases at the Earth's surface. Thus the temperature of the surface increases, thereby increasing the heat flux to the overlying troposphere, where convection transfers heat upward. For this reason, temperature increases not only at the Earth's surface but also in the entire troposphere. The magnitude of the warming, however, is determined such that the flux of outgoing radiation at the top of the atmosphere remains unchanged, despite the increase in the concentration of greenhouse gas in the atmosphere.

The warming is magnified further due to water vapor feedback as the absolute humidity of the air increases in the troposphere, keeping relative humidity unchanged. To evaluate quantitatively the influence of water vapor feedback upon the simulated warming, Manabe and Wetherald

(1967) conducted another set of runs in which that feedback was disabled. In these runs, the distribution of absolute humidity was prescribed to remain constant rather than being adjusted to maintain a constant relative humidity. From the differences among the three states of radiative-convective equilibrium thus obtained, they estimated the magnitude of the equilibrium response of surface temperature in the absence of water vapor feedback. They found that surface temperature increases or decreases by approximately 1.3°C in response to doubling or halving of the atmospheric CO_2 concentration, respectively. This change in temperature is substantially smaller than the values of 2.4°C and 2.3°C obtained in the presence of water vapor feedback. The results from these experiments indicate that water vapor exerts a strong positive feedback effect, which magnifies the surface temperature change by a factor of ~1.8.

In sharp contrast to the Earth's surface and troposphere, where temperature increases in response to a doubling of the atmospheric CO_2 concentration, cooling occurs in the stratosphere as indicated in figure 3.5. Because of the absence of convective heating, the heat balance of the stratosphere is maintained radiatively between the heating due to absorption of solar radiation by ozone and the net cooling due to emission and absorption of longwave radiation mainly by carbon dioxide, as shown by figure 3.3b. If the atmospheric CO_2 concentration doubles, for example, the longwave cooling intensifies and temperature in the stratosphere equilibrates at a lower value. This is the main reason why the stratospheric cooling occurs in the CO_2-doubling experiment, as shown in figure 3.5.

The cooling of the stratosphere results in a reduction of the outgoing longwave radiation from the top of the atmosphere. Because the model stratosphere is in local radiative equilibrium, the reduction of the outgoing longwave radiation at the top of the atmosphere is equal to that of the net upward flux of longwave radiation at the tropopause, which is the interface between the troposphere and stratosphere. Thus, the warming of the surface-troposphere system is larger than it would be in the absence of stratospheric cooling. In short, the cooling of the stratosphere results in an increase in the warming of the surface-troposphere system, as pointed out, for example, by Hansen et al. (1984).

Global mean temperature in the stratosphere has been decreasing during the past several decades. According to figure 3.6, which shows the time series of global mean temperature obtained from satellite microwave sounding and radiosonde observations, the global mean temperature has decreased at the rate of ~0.4°C per decade in the lower stratosphere during the past half-century, in contrast to the warming of ~0.2°C per decade in the lower troposphere during the same period. The reversal of temperature trends

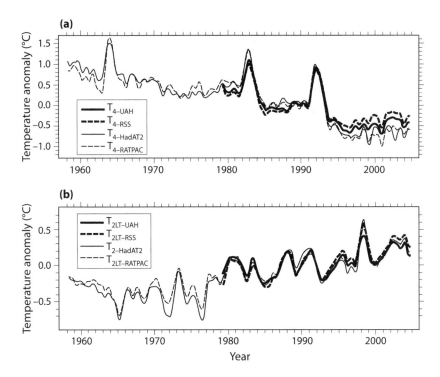

FIGURE 3.6 Time series of observed temperature anomalies (*a*) in the lower stratosphere and (*b*) in the lower troposphere obtained by Karl et al. (2006) from the analysis of satellite microwave soundings (UAH, RSS, and VG2) and that of radiosonde observations (UKMO HadAT2, and NOAA RATPAC). In their analysis, the lower stratosphere is a thick layer between 10 and 30 km, and the lower troposphere is the layer below 6 km. The two time series are monthly mean anomalies relative to the period 1979–97 and are smoothed with a seven-month running mean filter. For identification of the satellite microwave sounding and radiosonde observation acronyms, see IPCC (2007). From Trenberth et al. (2007).

between the stratosphere and troposphere appears to be in qualitative agreement with the result obtained from the radiative-convective model described here. It is likely, however, that the cooling of the stratosphere is attributable not only to the increase in the concentration of carbon dioxide but also to the reduction of ozone in the stratosphere, as noted, for example, by Ramaswamy (2006). For further discussion of this subject, see chapter 1 of the special report by the Intergovernmental Panel on Climate Change and the Technical and Economic assessment Panel (IPCC/TEAP, 2005).

As discussed in the comprehensive review conducted by Ramanathan and Coakley (1978), for example, 1-D radiative-convective models have been very useful for obtaining a preliminary estimate of the change in the global mean temperature in the atmosphere and at the Earth's surface as the concentration of greenhouse gases changes. Furthermore, the construction of the 1-D model turned out to be a critically important step toward the development of 3-D general circulation models of the coupled atmosphere-ocean-land system, which have become indispensable for exploring the climate change of not only the industrial present but also the geologic past. In the following chapter, we shall briefly describe the early development of general circulation models of the atmosphere and evaluate their performance in simulating the global distribution of climate.

CHAPTER 4

General Circulation Models

General circulation models (GCMs) of the atmosphere evolved from the dynamical models of numerical weather prediction that have become indispensable for making the weather forecasts that are widely used in our daily lives. In the 1950s Norman Phillips (1956) made the first attempt to simulate the general circulation of the atmosphere at the Institute for Advanced Study in Princeton, using a simple model that he developed there for numerical weather prediction. Here we describe briefly this truly pioneering study.

In the model developed by Phillips, the atmosphere consisted of two layers. In both the upper and lower layers, wind vectors and temperatures were determined at a regularly spaced two-dimensional (2-D) array of grid points. At each grid point, the model computed the temporal variations of wind and temperature by numerically integrating the equations of motion and the thermodynamic energy equation, respectively. The computational domain of the model was an idealized zonal channel with latitudinal width of 10,000 km.

Starting from an isothermal atmosphere at rest, Phillips conducted a numerical time integration of the model, in which the atmosphere was heated in low latitudes and cooled in high latitudes, with friction imposed at the Earth's surface. Because of the thermal forcing, the latitudinal gradient of temperature increased gradually with time. Meanwhile, very broad westerly winds developed in the atmosphere along with an overturning circulation in the meridional direction, with rising motion south of the center of the channel and sinking motion north of the center. Toward the end of the first stage of the experiment, which lasted about 30 days, the vertical shear of the zonal wind exceeded the critical value for the development of large-scale disturbances. At that point in the integration, small perturbations introduced into the calculation caused the rapid development of planetary waves with wavelengths of several thousand kilometers. Although nonlinear computational instability prevented the flow from reaching a statistically steady state, the model yielded successfully the intense westerly jet and associated planetary waves similar to those observed in the middle latitudes. Under the influence of the planetary wave disturbances, the meridional overturning circulation mentioned above was essentially confined to low latitudes, resembling the observed Hadley circulation, which features rising motion in the tropics and sinking motion in the subtropics.

Encouraged by the pioneering attempt of Phillips to simulate the salient features of the atmospheric circulation that control the global distribution of climate, several groups began to develop GCMs at various institutions such as GFDL (Smagorinsky, 1963); Lawrence Livermore National Laboratory (Leith, 1965); the Department of Meteorology at the University of California, Los Angeles (UCLA; Mintz, 1965); and the National Center for Atmospheric Research (Kasahara and Washington, 1967). (See chapter 7 of Edwards [2010] for a comprehensive account of these developments.) For the convenience of the discussion in the following sections, we describe here the structure and performance of the models developed at two of these institutions, UCLA and GFDL.

The UCLA Model

At the UCLA Department of Meteorology, Yale Mintz and Akio Arakawa constructed a global two-layer model of the troposphere with realistic distributions of land, sea, and the elevation of continental surfaces (Mintz, 1965, 1968). As in the model developed by Phillips, the UCLA model predicted wind and temperature at regularly spaced grid points in the upper and lower layers of the troposphere. Starting from an isothermal atmosphere at rest, a numerical time integration of the model was performed over the period of a few hundred days. Throughout the course of this integration, the incoming solar radiation during the month of January was prescribed at the top of the atmosphere. In addition, the geographic distribution of January monthly mean sea surface temperature was prescribed based on observations. The temperature at the continental surface was not prescribed, but instead was calculated by the model such that it satisfied the requirement of heat balance at the surface. Toward the end of the integration, the model reached a quasi-stationary state in which there was a balance between the incoming solar radiation and outgoing longwave radiation at the top of the atmosphere. Although the computational resolution of the model was coarse by today's standards, with a grid spacing of

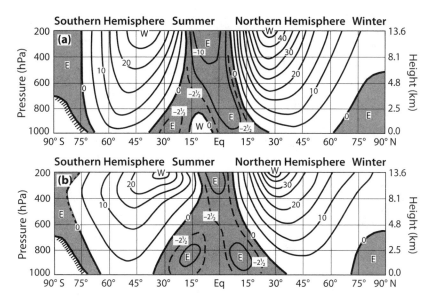

FIGURE 4.1 Latitude-height profile of zonal wind (m s⁻¹, positive eastward), averaged over the entire latitude circle for the month of January: (*a*) simulated; (*b*) observed. From Mintz (1965).

9° in longitude and 7° in latitude, the model simulated well not only the time-mean flow but also the amplitude of the transient wave disturbances observed in the atmosphere. The successful simulation was attributable in no small part to the finite-difference formulation of the so-called primitive equations of motion developed by Arakawa (1966). Owing to his very ingenious formulation, it was possible to perform stable numerical time integration of the model without the nonlinear computational instability encountered by Phillips. Here we present some highlights of the results they obtained.

Figure 4.1a presents a 2-D profile of the zonally averaged, zonal (i.e., west-to-east) component of wind obtained from the model described above. Intense westerly winds predominate in the upper troposphere of both hemispheres, with relatively weak easterlies (shaded) in low and high latitudes. The westerlies are substantially stronger in the Northern Hemisphere, where it is winter in January, than in the Southern Hemisphere in summer, in excellent agreement with observations (shown in figure 4.1b). Although the successful simulation of the zonal wind field may be attributable in no small part to the realistic prescription of the static stability of the troposphere and the spatial distribution of sea surface temperature, it is very impressive that the model simulated so successfully the latitude-height profile of zonal wind in the atmosphere.

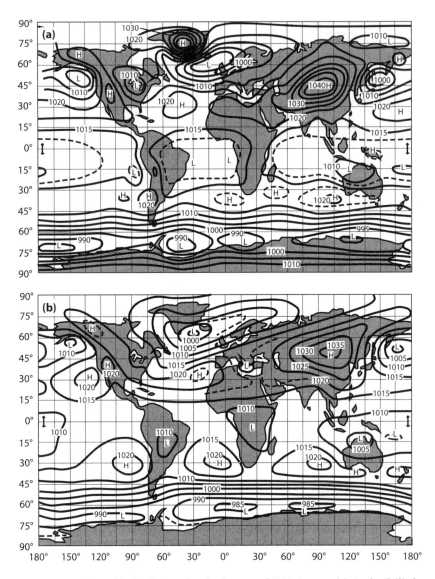

FIGURE 4.2 Geographic distribution of sea-level pressure (hPa) in January: (*a*) simulated; (*b*) observed. High and low pressure centers are designated H and L, respectively. From Mintz (1965).

The geographic distribution of sea-level pressure simulated by the model is compared with observations in figure 4.2. The model simulates well the large-scale pattern of monthly mean sea-level pressure observed in January. In the Northern Hemisphere, for example, the model places reasonably well the "Siberian high" in central Asia, the "Icelandic low" in the northern North Atlantic, and low-pressure centers in the North Pacific Ocean and high-pressure zones in the subtropics. In the Southern Hemisphere, the model simulates quite well the zonal belt of large meridional pressure gradient in the Southern Ocean, where intense westerly surface winds are maintained. Because wind vectors are almost parallel to the isobars, the successful simulation of sea-level pressure implies that the model simulates well the geographic distribution of wind in the near-surface layer of the atmosphere. In short, the model successfully reproduced the salient features of the atmospheric circulation such as the jet stream in the upper troposphere and the distribution of wind near the Earth's surface. The successful simulation was a major breakthrough in the development of atmospheric GCMs.

The GFDL Model

During the late 1950s and early 1960s, Smagorinsky (1958, 1963) also succeeded in the long-term integration of a two-layer model of the atmosphere. To prevent the computational instability that Phillips encountered in the time integration of his model, Smagorinsky used a nonlinear formulation of viscosity. Encouraged by the success of this approach, he began to develop a very ambitious plan for the construction of a comprehensive atmospheric GCM.

In early 1958, Smagorinsky invited Manabe, then just finishing his PhD at the University of Tokyo, to move to the United States and join his team in developing this GCM of the atmosphere. At that time, Manabe was having trouble finding work in postwar Japan, and he quickly accepted—making what would turn out to be one of the most important decisions of a six-decade-long career. By the fall, he had joined Smagorinsky's group in the general circulation research section of the United States Weather Bureau, located in a suburb of Washington, DC.

On a beautiful sunny day in the fall of 1958, Smagorinsky picked up Manabe at Washington National Airport, just across the Potomac River from the city. As soon as they arrived at his home, Smagorinsky began to talk about his very ambitious plan for the development of a comprehensive GCM that would explicitly incorporate the dynamics of general circulation, radiative heat transfer, and the hydrologic cycle. It was a blueprint for the development of the type of climate model that has since become indispensable for predicting climate change. Inspired by the plan, Manabe immersed himself deeply in this very challenging project.

Realizing that Smagorinsky was making excellent progress in developing the dynamic component of the model, he focused his attention on the other components, such as radiative heat transfer, moist and dry convection in the atmosphere, and the heat and water budgets at the continental surface. Smagorinsky had invited Professor Fritz Möller, an expert in radiative heat transfer, to visit his research group, and Möller offered valuable advice. To overcome problems with numerical stability and computational efficiency, the parameterization of processes that could not be explicitly resolved by the model (so-called sub-grid-scale processes) was made as simple as possible. By the middle of the 1960s, a model was constructed that successfully combined the dynamical component with the other components mentioned above. The results obtained from this model-development effort were published in two papers (Manabe et al., 1965; Smagorinsky et al., 1965). The first paper described the results from the version of the model without the hydrologic cycle, and the second paper described the results from the version in which the hydrologic cycle was explicitly incorporated. In the following sections we describe the structure of the second version (hereafter, the "annual mean model") and present highlights of the results obtained from it.

The Annual Mean Model

This model specified not only wind vectors and temperature but also humidity, at a 3-D array of grid points that were ~500 km apart in the horizontal direction and at nine unevenly spaced levels between the Earth's surface and the middle stratosphere. The rates of change of the three variables were computed using the primitive equations of motion, thermodynamic energy equation, and continuity equation of water vapor, respectively. As shown in figure 4.3, the model explicitly incorporated various physical processes such as thermal advection by large-scale circulation, solar radiation, longwave radiation, moist and dry convection, heat of condensation, and so on. In addition, the model incorporated the contribution from the exchange of momentum, heat, and water vapor between the Earth's surface and the overlying atmosphere through surface friction, sensible heat flux, and evaporation.

To compute the vertical distributions of solar and terrestrial radiation, the model used the method developed for the 1-D radiative model described in chapter 3. Given the distributions of not only temperature but also CO_2 and clouds, it computed the rate of temperature change due to the absorption of solar radiation and emission and absorption of longwave radiation in the atmosphere, using the equation of radiative transfer.

One of the important processes that controls the vertical distribution of temperature and that of water vapor is deep moist convection that penetrates into the upper troposphere. A conceptual model of deep convection is

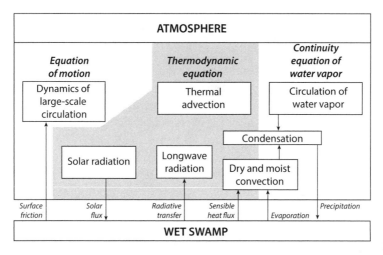

FIGURE 4.3 Box diagram illustrating the processes included in the initial version of the GFDL model. From Manabe et al. (1965).

the "giant hot tower" hypothesis developed by Riehl and Malkus (1958). In the hot towers that often develop in strong convective rainstorms, intense updrafts of saturated air are partially compensated mass-wise by intense downdrafts of saturated air, maintaining the moist adiabatic temperature profile that follows the ascent of a saturated air parcel in the atmosphere. Because of its huge size, the entrainment of relatively dry air from the surrounding region is not very effective in reducing humidity inside the tower. For this reason, a giant hot tower often extends well beyond the lower troposphere into the upper troposphere, often reaching the tropopause. This contrasts with relatively shallow convection such as trade-wind cumulus, in which an intense updraft of saturated air is compensated locally by a surrounding slow sinking of dry air. It should be noted that mesoscale convective systems with hot-tower characteristics often develop not only in low latitudes but also in middle latitudes, as pointed out, for example, by Zipser (2003). Examples include the eye wall of tropical cyclones, squall lines, and frontal systems that yield rapidly fluctuating, torrential rainfall.

In the model introduced here, deep convection is represented by the procedure called "moist convective adjustment," proposed by Manabe et al. (1965). This neutralizes the vertical gradient of temperature in a saturated air column whenever it exceeds the moist adiabatic lapse rate (i.e., the vertical gradient of temperature obtained by raising a saturated air parcel adiabatically). Moist convective adjustment was developed originally to prevent the computational instability that occurs in a saturated layer with moist adiabatically unstable stratification. Nevertheless, it mimics quite well deep mesoscale convection, in which an updraft of saturated air is compensated by a downdraft of saturated air, as in a hot tower as described above. Moist convective adjustment plays a critically important role in controlling the static stability of the entire model troposphere in low latitudes.

For the sake of simplicity, many idealizations were made in the construction of the initial version of the model. For example, solar radiation did not vary with season. Instead, annually averaged latitude-dependent insolation was prescribed at the top of the atmosphere. The model also had idealized geography: the domain was limited to a single hemisphere with a free-slip boundary placed at the equator. The Earth's surface was flat and horizontally uniform, covered by a swamp-like, wet surface with unlimited availability of water but without heat capacity. The temperature of the surface was determined such that it satisfied its heat balance, implicitly assuming that the heat exchange between the Earth's surface and its interior was absent.

Starting from the initial condition of an isothermal and dry atmosphere at rest, the numerical time integration of the model was performed over a period of 187 days. By the 150th day, the temperature in the atmosphere and at the Earth's surface ceased to change systematically. Figure 4.4a shows the latitude-height profile of zonal mean temperature averaged over the last 30 days of the time integration; figure 4.4b shows the distribution of observed temperature. Comparing the two, one finds that the two profiles do resemble each other. For example, the model reproduces, albeit crudely, the variation with latitude of the height of the tropopause, which is the interface between the convective troposphere and overlying stable stratosphere. Thus, it appears likely that this model contained the basic physical processes necessary for the formation and maintenance of the tropopause.

Although it is not shown here, the model also qualitatively reproduced the regions of enhanced precipitation associated with the intertropical convergence zone (ITCZ) in the tropics and the extratropical storm tracks of middle latitudes, as well as the belt of minimum precipitation in the subtropics. The meridional span of the tropical rain belt was too narrow, however, not only because of the free-slip wall artificially placed at the equator but also owing to the absence of seasonal variation in the model. The absence of seasonal variation also accounts for the failure of the tropical tropopause to extend far enough toward the middle latitudes, as shown in figure 4.4a.

The Seasonal Model

Because of its encouraging performance in simulating the latitude-height profile of zonal mean temperature in the atmosphere, the annual mean model was converted to a global model with realistic geography and

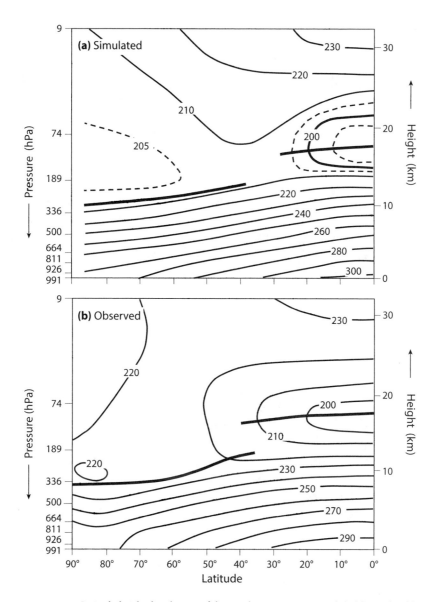

FIGURE 4.4 Latitude-height distribution of the zonal mean temperature (K): (*a*) simulated by the initial version of the GFDL model; (*b*) observed. The thick solid line denotes the height of the tropopause. From Manabe et al. (1965).

seasonal variation of incoming solar radiation. At the oceanic surface, the horizontal distribution of seasonally varying sea surface temperature was prescribed as observed. The temperature of the continental surface was determined such that it satisfied the balance requirement between the heating due to absorption of solar radiation and the cooling due to the net upward fluxes of longwave radiation, sensible heat, and the latent heat of evaporation. The evaporation rate was determined as a function of surface temperature and soil moisture (i.e., the amount of water contained in the root zone). The temporal variation of soil moisture was computed based on a water balance requirement, with rainfall and snowmelt adding water to the soil and runoff and evaporation removing it. The water-equivalent depth of snow was predicted from the difference between the gain due to snowfall and the loss due to sublimation and melting.

With the availability of increasingly powerful computers, it was possible to double the computational resolution of the model, reducing the grid interval from ~500 to ~250 km. But, in order to save computer time, a low-resolution version was used to initialize the model. The top-of-atmosphere flux of incoming solar radiation and sea surface temperature in January were imposed as boundary conditions for this preliminary run (Holloway and Manabe, 1971). Once the initial state had been determined, the numerical time integration of the seasonal model was continued for a period of about 3.5 years. Although one of the fastest computers available in the early 1970s was used for this simulation, it took several thousand hours of computer time to complete the time integration. The output from the last two annual cycles of the simulation were used for the analysis presented here.

Figure 4.5a shows the geographic distribution of the annual mean precipitation rate obtained from the seasonal model. In the tropics, a large area of intense rainfall occupies the western Pacific and Indian Oceans. Another area of intense rainfall is located in the Amazon basin. In the eastern tropical Pacific, the northern branch of the tropical rain belt is located just north of the equator, whereas the southern branch that originates from the central Pacific extends southeastward. The geographic pattern of tropical precipitation obtained from the model is in reasonably good agreement with the observed pattern shown in figure 4.5b.

In low latitudes, zonal mean precipitation is strongly controlled by the Hadley circulation. Rising motion near the equator accounts for the intensity of the precipitation in the deep tropics, whereas sinking motion in the subtropics inhibits precipitation there. This accounts for the meager precipitation in regions such as the Sahara and Australia, in agreement with observations. In middle latitudes, the rate of precipitation is relatively large, again in both the model and observations, and is attributable mainly to the frequent passage of extratropical cyclones that often yield heavy precipitation.

Comparing the simulated distribution of precipitation with that of sea surface temperature (not shown) in the tropics, one can see that the rate of precipitation in low latitudes is usually large in those regions where the sea

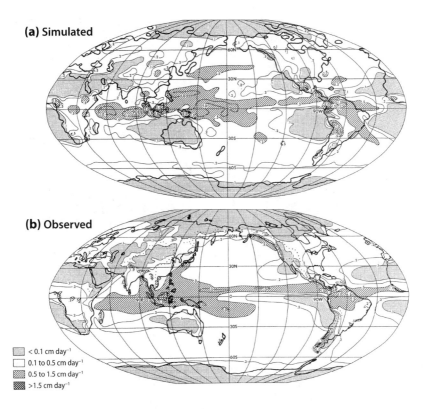

(a) Simulated

(b) Observed

■ < 0.1 cm day⁻¹
□ 0.1 to 0.5 cm day⁻¹
▨ 0.5 to 1.5 cm day⁻¹
▨ >1.5 cm day⁻¹

FIGURE 4.5 Global distribution of the annual mean rate of precipitation (cm day⁻¹): (*a*) simulated; (*b*) observed. Note that the contours of observed precipitation are hand-smoothed over the continents and are different in details from the original map. From Manabe and Holloway (1975).

surface is warmer than surrounding areas, in agreement with observations. This is because tropical disturbances, including tropical cyclones, often develop over such regions, yielding heavy precipitation. As explained by Manabe et al. (1970, 1974), deep convection, represented in the model by moist convective adjustment, is responsible for the formation of a warm core in the upper troposphere, playing a critically important role in the development of tropical cyclones.

It is notable that the model simulated quite well not only the geographic distribution of rainfall but also that of atmospheric circulation, particularly in low latitudes, where the Coriolis force is weak and moist convection predominates. Figure 4.6 illustrates the spatial distributions of monthly mean flow near the Earth's surface in January (figure 4.6a) and July (figure 4.6b). In both months, the flow fields are characterized by divergence from oceanic anticyclones in the subtropics and convergence into a narrow belt

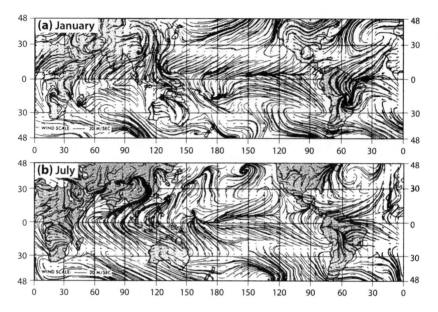

FIGURE 4.6 Airflows (indicated by streamlines and wind vectors) near the Earth's surface, simulated (*a*) for January and (*b*) for July. From Manabe et al. (1974).

near the equator—that is, the ITCZ. In the eastern Pacific of the model, for example, the ITCZ is located to the north of the equator in both July and January. However, it is located farther north in July than in January. Over the Atlantic, the ITCZ undergoes a similar movement with respect to season, though it approaches very close to the equator around January. The seasonal movement of the ITCZ described above corresponds well with that of the tropical rain belt that forms in those parts of the deep tropics where sea surface temperature is warmer than the surrounding areas, as noted above. The results presented here indicate that successful simulation of the distribution of tropical precipitation and of the flow field is attributable in no small part to the realistic distribution of temperature prescribed at the oceanic surface.

One of the most striking features of the near-surface flow field simulated by the model is the intense cross-equatorial southerly wind over the Indian Ocean and western side of the Pacific Ocean in July. The southerly wind supplies a large amount of moisture to the monsoon rains over India and Southeast Asia. In January, the direction of surface flow is reversed in the northern Indian Ocean, with the northerly wind blowing out of the Siberian high, reaching the equator, and entering the ITCZ slightly to the north of the equator. The features of the surface flow described above are in excellent agreement with those in the actual atmosphere, shown in figure 4.7.

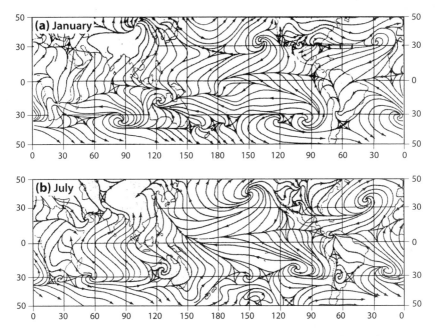

FIGURE 4.7 Observed airflows (indicated by streamlines) at the Earth's surface, (*a*) for January and (*b*) for July.

In this chapter, we have presented an overview of the early development of GCMs of the atmosphere, evaluating their performance in simulating the distributions of wind, temperature, and precipitation. The successful simulations provide ample evidence that a GCM is potentially a very powerful tool for studying and predicting climatic change. In the following chapters, we shall introduce GCMs with increasing complexity and describe how they have been used in order to elucidate the physical mechanisms of global warming and associated changes in the water cycle of this planet.

CHAPTER 5

Early Numerical Experiments

Polar Amplification

The use of GCMs to study changes in climate began at GFDL in the late 1960s and early 1970s. The results of this work were described in two companion papers published in the mid-1970s (Manabe and Wetherald, 1975; Wetherald and Manabe, 1975). In the former study, the total response of climate to a hypothetical doubling of the atmospheric CO_2 concentration was investigated. The response to a 2% change in solar irradiance was evaluated in the latter study. Although the model used for these studies was similar to the annual mean model described in the preceding chapter, there was one important difference. To incorporate the positive feedback effect of water vapor, the model used the distribution of water vapor determined by the model for the computation of radiative transfer—in contrast to the original annual mean model, which used the observed distribution.

Because of limited computing power available at that time, it was desirable to minimize the computational requirements of each numerical experiment. To achieve this goal, the spatial domain of the model was reduced from one hemisphere to a single sector that covered only one-sixth of the globe. This sector was bounded by two meridians separated by 120° of longitude and extended from the equator to 81.7°, as shown in figure 5.1. In the atmosphere, cyclic continuity was imposed at the eastern and western boundaries such that atmospheric disturbances that exited through one of the north–south boundaries would re-enter the domain through the other boundary. Insulated free-slip walls were placed at the equator and at 81.7°. Despite the reduced spatial domain, the sector was designed to be wide enough to maintain the planetary waves that play a critically important role in controlling the dynamics of the atmospheric circulation.

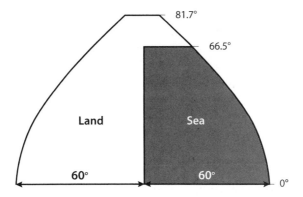

FIGURE 5.1 Computational domain of the general circulation model used by Manabe and Wetherald (1975).

This sector domain was subdivided further into land and sea, each of which had a longitudinal span of 60°. In contrast to previous models, which prescribed sea surface temperatures based on observations, the temperatures of both the oceanic and continental surfaces were determined by a heat balance requirement, implicitly assuming that the heat capacity of the sea surface is zero, as it is at the continental surface. This assumption enabled temperatures in the model to respond to the changes in top-of-atmosphere energy balance that result from the imposed changes in CO_2 concentration or solar irradiance. When the sea surface temperature dipped below the freezing point of seawater (−2°C), the sea surface was assumed to be covered by sea ice with a higher albedo. Although the oceanic surface of the model was prescribed to be always wet with an unlimited availability of water, soil moisture and snow depth were computed at the continental surface as in the annual mean model described in chapter 4. By allowing the model to predict the coverage of highly reflective sea ice and snow, the model was able to simulate the albedo feedback that enhances the sensitivity of climate.

The CO_2-Doubling Experiment

To estimate the equilibrium response of temperature to a doubling of the CO_2 concentration in the atmosphere, two sets of numerical time integrations of the model were performed. In the first set of integrations, the atmospheric CO_2 concentration was prescribed at 300 ppmv. Two 800-day integrations were run, differing only in their initial conditions. Although both started from an isothermal and dry atmosphere at rest, one initial condition was very warm and the other was very cold. Despite these large differences in temperature at the start of the two runs, the temperature of the atmosphere was practically identical toward the end of the time integrations. A quasi-equilibrium state was obtained, averaging the two 100-day mean states at the end of the two runs. An analogous set of simulations was made with the atmospheric CO_2 concentration prescribed at 600 ppmv, and a quasi-equilibrium state obtained in the same manner. The equilibrium response to CO_2 doubling was then determined from the difference between these two quasi-equilibrium states.

Figure 5.2a illustrates the equilibrium response of zonal mean temperature to the imposed CO_2 doubling. In agreement with the results from the 1-D radiative-convective model described in chapter 3, temperature increased not only at the Earth's surface but also in the troposphere, whereas it decreased in the stratosphere. In the lower troposphere below a height of ~5 km, the magnitude of the warming increased with increasing latitude and was particularly large in the near-surface layer of the atmosphere, owing mainly to the poleward retreat of snow and sea ice that reflects a large fraction of incoming solar radiation. A close inspection of figure 5.2 reveals, however, that the large warming in high latitudes was not confined to the near-surface layer of the atmosphere. Instead, it extended upward to the mid-troposphere. This underscores the possibility that the polar amplification of the warming is attributable not only to the positive albedo feedback but also to a reduction of tropospheric static stability in high latitudes due to the change in the large-scale atmospheric circulation, as pointed out, for example, by Held (1978).

In contrast to high latitudes, where the magnitude of the warming decreased with increasing height in the troposphere, it increased with increasing height in low latitudes. This is attributable mainly to deep moist convection that often develops in the tropics, which keeps the vertical temperature gradient near the moist adiabatic lapse rate. Because the moist adiabatic lapse rate decreases with increasing temperature, the magnitude of the warming is larger in the upper troposphere than in the lower troposphere, as shown in figure 5.2. According to microwave soundings from satellites (Fu et al., 2004; Fu and Johanson, 2005), the temperature of the tropics has increased more in the upper troposphere than at the Earth's surface during the past several decades, in agreement with the results presented here. It is interesting to note that the area-mean static stability of the model troposphere hardly changed in response to CO_2 doubling because of the compensation between high and low latitudes. Thus, the assumption of invariant static stability of the troposphere that was employed in the

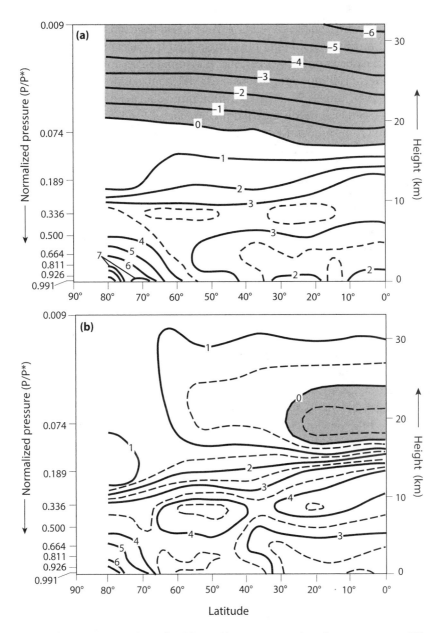

FIGURE 5.2 Latitude-height profile of the equilibrium response of zonal mean temperature (°C) in the atmosphere. (*a*) Response to a doubling of the atmospheric concentration of carbon dioxide (Manabe and Wetherald, 1975); (*b*) response to a 2% increase in the solar irradiance (Wetherald and Manabe, 1975). *P*, pressure; *P**, surface pressure.

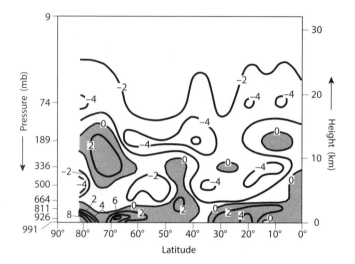

FIGURE 5.3 Latitude-height profile of the change in zonal mean relative humidity in response to the doubling of the atmospheric concentration of carbon dioxide. Zonal mean relative humidity is the percentage of the change in zonal mean vapor pressure relative to the zonal mean saturation vapor pressure. From Manabe and Wetherald (1975).

globally averaged radiative-convective model of the atmosphere introduced in chapter 3 appears to have been justified.

Averaged over the entire model domain, temperature increased by 2.9°C at the Earth's surface, which is somewhat larger than the 2.4°C obtained from the radiative-convective model with water vapor feedback activated. The 3-D model described here incorporated not only the water vapor feedback but also the albedo feedback of snow and sea ice. It is therefore likely that the difference in the magnitude of the warming between the two models is mainly attributable to the albedo feedback.

To examine how the water vapor feedback operates in the model, figure 5.3 illustrates the change in zonal mean relative humidity in response to the doubling of the CO_2 concentration in the atmosphere. As this figure shows, the distribution is very patchy because the magnitude of the systematic change of relative humidity is much smaller than that of its natural variability. Nevertheless, one can identify some systematic changes in the distribution of relative humidity. For example, relative humidity increases by a few percent in the lower troposphere below 700 hPa, whereas it decreases by a few percent in the upper troposphere between 300 and 700 hPa. Referring to the study of Manabe and Wetherald (1967) that evaluated the relationship between the relative humidity of the troposphere and mean surface temperature, one can estimate crudely the contribution

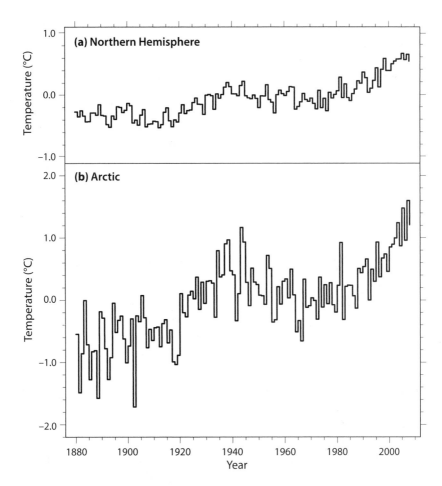

FIGURE 5.4 Time series of the mean surface temperature anomaly relative to the reference period 1880–1960 in (*a*) the Northern Hemisphere (0°–85° N) and (*b*) the Arctic region (65°–85° N). The time series obtained by Kelly et al. (1982) are extended to 2010 using data updated by Brohan et al. (2006).

of this change in relative humidity to the change in surface temperature. Our estimate suggests that it is no more than 0.1°C because of the partial cancellation between the contributions from the upper and lower troposphere. In short, the strength of the water vapor feedback in this 3-D model is likely to be similar to that in the 1-D radiative-convective model with fixed relative humidity described in chapter 3.

The polar amplification of global warming has been confirmed through climate observations conducted over the past 130 years. In figure 5.4, the time series of the annual mean surface temperature anomaly averaged over the Arctic region and the entire Northern Hemisphere are compared. This figure shows that both time series exhibit systematic warming trends with fluctuations on interannual, decadal, and multidecadal time scales. The average rate of the warming, however, is much larger in the Arctic than in the Northern Hemisphere as a whole, providing convincing evidence for the polar amplification of global warming. Although it is not shown here, the polar amplification of surface warming is not so evident in the Southern Hemisphere. This is because the deep vertical mixing of heat that predominates in the Southern Ocean has kept the magnitude of warming at the oceanic surface very small. This mechanism will be explored further in chapter 8, using a model that couples the atmosphere with the full oceans.

Change in Solar Irradiance

So far, we have presented results from a numerical experiment in which the atmospheric CO_2 concentration was doubled. Similar numerical experiments were also performed in which the solar irradiance (i.e., the average incoming solar radiation at the top of the atmosphere) was changed by +2%, −2%, and −4%, respectively. Figure 5.2b shows the latitude-height profile of the equilibrium response of zonal mean temperature to the +2% change in irradiance. Comparing it with the response to CO_2 doubling depicted in figure 5.2a, one finds that the responses of zonal mean temperature are quite similar in the troposphere below 15 km. For example, in the near-surface layer of the troposphere, the magnitude of the warming increases with increasing latitude and is at a maximum in high latitudes. On the other hand, in the upper troposphere around a height of 10 km, the warming decreases slightly with increasing latitude. This similarity is noteworthy when one considers the differences in thermal forcing between the +2%-solar-irradiance and doubled-CO_2 cases. Although both thermal forcings are positive, the former decreases with increasing latitude at a much larger rate than the latter. In view of the difference in the latitudinal gradient of the thermal forcing, it is surprising that the latitudinal profile of zonal mean temperature change is quite similar between the two types of thermal forcing.

The similarity between the responses to increased solar irradiance and increased CO_2 has been found in experiments using a different climate model. Hansen et al. (1984) conducted a similar pair of experiments using a GCM developed at the Goddard Institute for Space Studies (GISS) of the National Aeronautics and Space Administration (NASA). They found that latitude-height profile of the zonal mean response was practically identical between the +2%-solar-irradiance and CO_2-doubling experiments, in agreement with the results presented here. These studies suggest that, in the troposphere and at the Earth's surface, the latitudinal profile of temperature change has little dependence upon the profile of the thermal forcing. Instead, it depends essentially upon the magnitude of the total thermal forcing that controls the polar amplification of near-surface temperature. These results also suggest that, in the troposphere and at the Earth's surface, not only the +2% irradiance, but also the −2% and −4% experiments may be used as surrogates for the CO_2-doubling, CO_2-halving, and CO_2-quartering experiments, respectively.

The similarity between the CO_2-doubling and +2%-irradiance profiles suggests the possible existence of a dynamical mechanism that prevents the meridional temperature gradient from exceeding a certain critical value in the troposphere, where the static stability hardly changes. Such a mechanism has been proposed, for example, by Smagorinsky (1963), based upon the analysis of the long-term integration of his two-layer model mentioned in chapter 4. Here we describe briefly his hypothesis, which has been discussed further by Stone (1978). Suppose that the meridional temperature gradient exceeds the critical value for the instability of the midlatitude westerlies (i.e., baroclinic instability). As a result, planetary waves would be expected to amplify, enhancing the poleward transport of heat and preventing a further increase in the meridional temperature gradient in the troposphere. One can speculate that a similar mechanism is in operation in the model troposphere in both the CO_2-doubling and +2% experiments. This may be an important reason why the latitudinal distribution of zonal mean temperature change is similar between the two experiments, even though the distribution of zonal mean thermal forcing is quite different.

The latitudinal profiles of zonal mean surface temperature for four different values of solar irradiance are shown in figure 5.5. In addition to the profile from the control run, using a solar irradiance value of 340 W m^{-2}, the figure includes the profiles from three additional runs in which the solar constant is changed by +2%, −2%, and −4%, respectively. This figure shows that the meridional temperature gradient increases from the +2% run to the −4% run as the temperature decreases and the albedo feedback of snow and sea ice intensifies. Table 5.1 tabulates the equilibrium responses of the area-mean surface temperature to +2%, −2%, and −4% changes in the irradiance. For the sake of comparison, it also lists the responses obtained from the radiative-convective model introduced in chapter 3. Compared with the control run, the model warms by 3.04°C in response to a 2% increase in solar irradiance. This warming is substantially smaller than the 4.37°C cooling in response to a 2% reduction in irradiance and the 5.71°C cooling that occurs when the irradiance is further reduced from

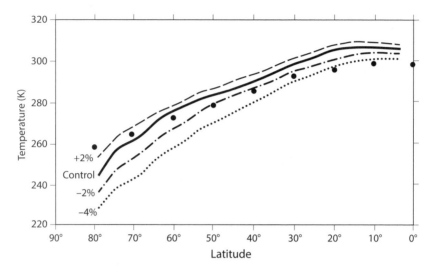

FIGURE 5.5 Latitudinal profiles of the zonal mean surface air temperature (K) obtained for four values of total solar irradiance: standard value (control; 1395.6 W m⁻²), +2%, −2%, and −4%. The observed annually averaged zonal mean surface temperature is indicated by black dots. From Wetherald and Manabe (1975).

TABLE 5.1 *Equilibrium response of mean surface air temperature to change in solar irradiance*

Change in Solar Irradiance	Change in Surface Temperature	
	GC Model	RC Model
From control to +2%	+3.04°C	+2.57°C
From control to −2%	−4.37°C	−2.55°C
From −2% to −4%	−5.71°C	−2.54°C

GC model, general circulation model introduced here; RC model, radiative-convective model described in chapter 3. From Wetherald and Manabe (1975).

−2% to −4%. In other words, the magnitude of the response to each 2% reduction in the solar constant increases with decreasing solar irradiance.

This dependence of the magnitude of the thermal response on global mean temperature in the GCM is dramatically different from the behavior of the radiative-convective model, where the responses to each 2% increment in solar irradiance are nearly identical at 2.57°C, 2.55°C, and 2.54°C. The difference in behavior between the two models is attributable mainly to the albedo feedback of snow and sea ice. In short, the response to a given thermal forcing increases with decreasing surface temperature as the areal coverage of the polar cap of snow and sea ice increases. In other words, if an increment of decreased thermal forcing moves the snow and sea ice boundaries equatorward by a given distance, the area of additional high-albedo surface is proportional to the length of latitude circle that defines those boundaries. Thus, each increment of cooling produces a greater amount of ice albedo feedback.

Energy balance models are powerful tools for understanding the role of ice-albedo feedback. In their simplest form, such models are 1-D, with temperature varying only with latitude. The components of the energy balance in these models are outgoing longwave radiation, absorption and reflection of incoming solar radiation, and meridional heat transport by large-scale atmospheric circulation, all of which are assumed to depend upon the perturbations of zonal mean surface temperature. Budyko (1969) and Sellers (1969) pioneered the construction of such models and used them to study the response of the polar ice caps to changes in solar irradiance. Held and Suarez (1974) and North (1975a, b, 1981) built upon these pioneering studies by conducting mathematically rigorous analyses of model behavior that yielded valuable insights into the response of ice caps and surface temperature to changes in solar irradiance.

Despite their simplicity, 1-D energy balance models (EBMs) anticipated many of the findings from the GCMs introduced here. For example, EBMs indicated that the response of global mean surface temperature to a given change in incoming solar radiation increases with decreasing solar irradiance, as the albedo feedback intensifies owing to the expansion of ice cover from the polar regions toward middle latitudes. They also indicated that ice cover becomes unstable when the solar irradiance dips below a threshold and the polar ice cap grows beyond a critical latitude (the so-called "large-ice-cap instability"). Under such circumstances, the positive ice albedo feedback overwhelms the negative feedback due to the temperature dependence of the emission of longwave radiation and the meridional diffusion of heat. Thus, ice cover can extend all the way to the equator, yielding an ice-covered planet reminiscent of the "Snowball Earth" that appeared intermittently approximately 750 to 550 million years ago (Harland, 1964; Hoffman et al., 1998; Pierrehumbert et al., 2011). According to the EBM studies of Budyko and Sellers, a reduction in solar radiation of less than 2% would be enough to induce such ice-cap instability. This conclusion differed from the study of Wetherald and Manabe described earlier in this chapter (1975), which indicated that a reduction of solar irradiance by much more than 2% (or a reduction in CO_2 concentration of more than a factor of two) would be required for the ice-cap instability to occur.

The sensitivity of the ice caps to solar irradiance was the subject of an in-depth analysis, which used a two-level GCM of the atmosphere (Held et al., 1981). The model was simplified such that a large number of numerical experiments could be performed with minimal computational cost and a quasi-equilibrium state could be obtained with relative ease. Held et al. found that the response of their model to changes in solar irradiance differed substantially from the response of the diffusive energy balance model. When solar irradiance is decreased by a few percent, pushing the albedo gradient at the ice-cap boundary to a position equatorward of 60° latitude, the model climate becomes very sensitive to the solar irradiance value, but this sensitivity is not a sign that ice-cap instability is close at hand. Instead, as the solar irradiance is lowered further, the model climate becomes less rather than more sensitive. These relatively stable large-ice-cap states exist for a range of ~5% of solar irradiance, while the ice cap extends all the way to 20° latitude. For still lower values of the solar constant, large-ice-cap instability develops and ice covers the entire planet.

Similar results were produced in a two-level diffusive EBM, closely patterned after the GCM, by making the diffusion coefficient for heat vary with latitude. Choosing diffusivity with a pronounced meridional structure resembling that of the effective diffusivity of the GCM, these authors found that the ice-cap boundary is less sensitive to solar irradiance when it is in the region in which effective diffusivity of the atmosphere increases with decreasing latitudes. On the other hand, greater sensitivity is seen when the ice-cap boundary enters the region in which the diffusivity decreases with decreasing latitude. The results from these experiments indicate that the large-ice-cap instability found in the EBM with constant diffusivity would require a much larger decrease in solar irradiance to occur in the GCM. The dependence of effective diffusivity on latitude is responsible for this difference in response, indicating that the transport of energy by transient eddies cannot be adequately represented by diffusivity that is independent of latitude. Thus, some caution is required in interpreting the quantitative aspects of results obtained from EBMs in which the thermal diffusivity does not depend upon latitude.

Seasonal Variation

So far, we have presented a series of studies of climate change that used highly idealized models with annually averaged insolation. By the 1970s, when increasingly powerful computers became available at GFDL, the GCMs used for studies of climate change were made to resemble the real

world more closely by including realistic geography and seasonally varying insolation. One such model was used by Manabe and Stouffer (1979,
1980) and revealed how the polar amplification of global warming depends
upon season. We will briefly describe the structure of this model before
discussing the results that were obtained from it.

The model covers the entire globe and has a realistic geographic distribution of oceans and continents. As with the seasonal model described in
chapter 4, it consists of a GCM of the atmosphere and a heat- and water-
balance model of the continental surface. In contrast to the earlier model,
sea surface temperature is not prescribed according to observations. Instead, the ocean is represented as a vertically isothermal layer with a thickness of ~70 m. The temperature of this "mixed layer" is determined such
that it satisfies the surface heat budget; no horizontal transport of heat
occurs within the model ocean. The heat exchange between the ocean
mixed layer and the deeper ocean is neglected. Although this exchange is
not very important over short time scales, it substantially affects changes
in surface temperature over multidecadal and centennial time scales. The
role of the deep ocean in climate change will be the subject of discussion
in chapters 8 and 9.

An important factor that affects the heat budget of mixed-layer ocean
is sea ice. It often covers the oceanic surface in high latitudes, reflecting
a major fraction of incoming solar radiation. It also reduces the heat exchange between the mixed-layer ocean and the overlying atmosphere,
substantially affecting the heat balance of the mixed layer. In the model,
the change in sea ice thickness is computed from the budget of sea ice that
involves snowfall, sublimation and melting at the top of the sea ice, and
freezing (or melting) at the bottom of the sea ice, as illustrated in figure
5.6. Here the rate of freezing (or melting) at the bottom of the sea ice is
determined such that the temperature of the mixed layer of ocean remains
at its freezing point (−2°C) despite the heat conduction through the sea
ice. The temperature at the top surface of the sea ice is determined such
that the requirement of heat balance is satisfied among solar radiation,
longwave radiation, sensible heat flux, sublimation, and heat conduction
through the sea ice. When the temperature thus computed exceeds the
melting point of sea ice (0°C), melting occurs, keeping the temperature at
the melting point of sea ice.

Using the atmosphere/mixed-layer-ocean model described above,
Manabe and Stouffer attempted to simulate the seasonal variation of climate. Starting from an isothermal initial condition at rest, they numerically integrated the model over 10 annual cycles. Initially, the global mean
surface temperature of the model changed very rapidly. Toward the end of

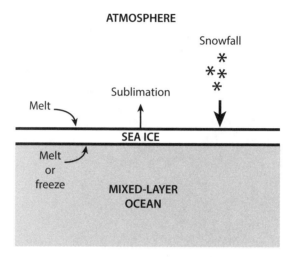

FIGURE 5.6 Components of the sea ice budget in the atmosphere/mixed-layer-ocean model constructed by Manabe and Stouffer (1979, 1980).

the integration, however, it closely approached the state of thermal equilibrium, in which the global mean flux of outgoing longwave radiation is practically identical to that of the net incoming solar radiation at the top of the atmosphere.

To evaluate the ability of the model to simulate the seasonal variation of surface temperature, maps of the difference in surface air temperature between August and February for the model were compared with observations (figure 5.7). Keeping in mind that the sign of the difference is opposite between the two hemispheres, one can use the magnitude of the difference as a rough indicator of the annual range of surface temperature variation.

Comparing figure 5.7a and b, one can see that the geographic distribution of the surface air temperature difference between August and February simulated by the model agrees well with observations. For example, the annual range of surface air temperature over oceanic regions is substantially smaller than the range over the continents, owing mainly to the thermal inertia of the mixed-layer ocean. It was encouraging to find that this relatively simple treatment of the ocean mixed layer allowed the model to simulate reasonably well the geographic distribution of the annual range of surface air temperature.

Using the atmosphere/mixed-layer-ocean model, Manabe and Stouffer performed a numerical experiment to investigate the climatic effect of increasing the atmospheric CO_2 concentration. The simulation described above, in which the atmospheric CO_2 concentration was held fixed at a

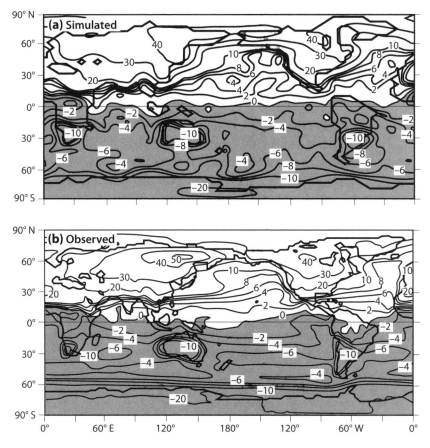

FIGURE 5.7 Geographic distribution of the difference in monthly mean surface air temperature (°C) between August and February: (*a*) simulated distribution; (*b*) observed distribution based upon data compiled by Crutcher and Meserve (1970) and Taljaad et al. (1969) for the Northern and Southern Hemispheres, respectively. Note that the contour interval is 2°C when the magnitude of the difference is less than 10°C, and is 10°C when the difference is more than 10°C. The area of negative difference is shaded. From Manabe and Stouffer (1980).

value of 300 ppmv, was used as a control run (1×C). To simulate a high-CO_2 climate, another run was made with the CO_2 concentration held fixed at four times the standard value (1200 ppmv; 4×C). This large change in CO_2 was used to magnify the response of climate that would be determined by comparing the simulated 4×C and 1×C climates. Because the radiative forcing of climate is proportional to the ratio of CO_2 concentration before and after the change, as explained in chapter 3, the climate change due to CO_2 quadrupling is approximately twice as large as the change due to CO_2 doubling.

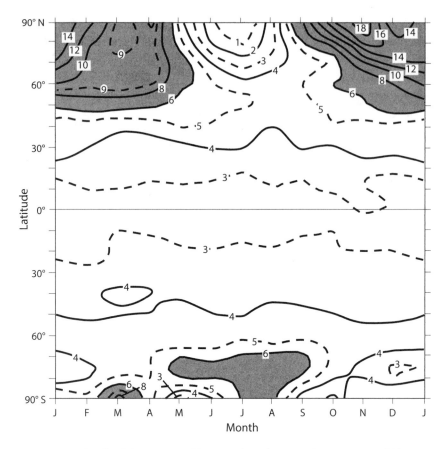

FIGURE 5.8 Simulated change in monthly averaged zonal mean surface temperature (°C) in response to the quadrupling of atmospheric CO_2 as a function of latitude and calendar month. From Manabe and Stouffer (1979, 1980).

The simulated change in zonal mean surface air temperature in response to CO_2 quadrupling is displayed as a function of latitude and calendar month, to illustrate the seasonal dependence of polar amplification of global warming, in figure 5.8. In high latitudes of the Northern Hemisphere, the magnitude of warming at the Earth's surface varies greatly with season. It is very large from fall to late spring, but is small in summer with no polar amplification. Although similar seasonal variation also appears near the Antarctic coast around 70° S, its amplitude is much smaller. Sea ice plays a key role in modulating the heat exchange between the water beneath sea ice and the overlying air, thereby controlling the seasonal dependence of the warming over the Arctic Ocean and its immediate vicinity.

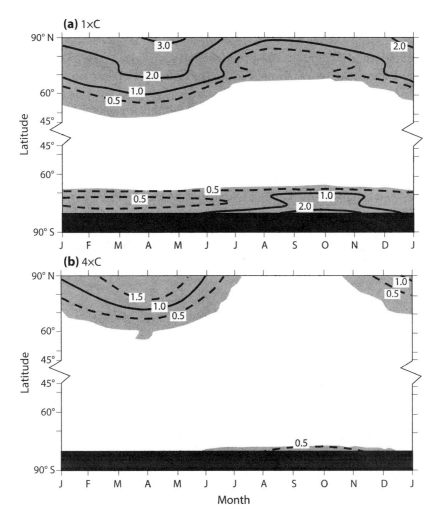

FIGURE 5.9 Simulated change in monthly averaged zonal mean sea ice thickness (m) as a function of latitude and calendar month: (*a*) control 1×C run; (*b*) 4×C run. Shading indicates the region where sea ice thickness exceeds 0.1 m. From Manabe and Stouffer (1980).

Figure 5.9a shows how the zonal mean thickness of sea ice in the control experiment (1×C) varies with season in the high latitudes of the two hemispheres. Poleward of 60° N, sea ice is thin and at a minimum during the three-month period of August–October. Sea ice thickness begins to increase in October and this increase continues until April, when it is at a maximum. While simulated surface air temperature decreases rapidly from summer to early winter as shown in figure 5.8, the temperature of the water beneath the sea ice remains at −2°C, the freezing point of seawater. Thus, the magnitude of the temperature gradient from the bottom to the top of the sea ice increases, enhancing the upward heat conduction that freezes water at the bottom of the sea ice. This process is responsible for the increase in sea ice thickness from September to April. Ice thickness begins to decrease rapidly during late spring owing to surface melting under increasing sunshine and warm air advected from the surrounding continents. Although similar seasonal variations of sea ice thickness occur near the Antarctic coast, their amplitude is small as compared with the Arctic Ocean.

Figure 5.9 indicates that the areal coverage of sea ice in the atmosphere/mixed-layer-ocean model is small in summer, when insolation is relatively strong, whereas it is large in winter, when insolation is relatively weak. This negative covariance between sea ice and insolation causes the total amount of reflected solar radiation during the annual cycle to be reduced owing to the seasonal variation of sea ice area. The consequent increase in absorbed solar radiation is one of the reasons why the mean surface air temperature was higher in the model with seasonal variation than in the model without such variation, as discussed by Wetherald and Manabe (1981). Comparing these simulations with and without annual variation, their results imply that the smaller the seasonal variation of incoming solar radiation, the stronger the albedo feedback and the lower the mean temperature at the Earth's surface. This relationship would lend support for the so-called astronomical theory of the ice ages, in which the growth of Northern Hemisphere ice sheets is facilitated by reduced summer insolation. For further discussion of this subject, see Hays et al. (1976) and Imbrie and Imbrie (1979, 1980).

Comparing the latitude/calendar-month distributions of zonal mean sea ice thickness obtained from the 4×C and 1×C runs (figure 5.9), one can see that the coverage and thickness of sea ice are reduced markedly in the 4×C run. The reduction of sea ice thickness is attributable in no small part to the increase in the absorption of incoming solar radiation by the mixed-layer ocean in summer, when sea ice coverage is at a minimum and is reduced markedly from the 1×C run. The increase in the absorption of solar energy during summer delays the growth of sea ice during the cold season and contributes to a reduction of sea ice thickness throughout a year.

This overall reduction of sea ice thickness contributes strongly to the large seasonal contrast in the magnitude of Arctic warming depicted in figure 5.8. In winter, for example, temperature is much colder at the top of the sea ice than at the bottom, where it remains at the freezing point of seawater. Because the rate of heat conduction is inversely proportional to sea ice thickness, the heat conducted through the ice increases as sea ice thins. This is the main reason why surface temperature increases over the Arctic

Ocean in response to the increase to CO_2 concentration. The magnitude of the warming is particularly large in winter, when the temperature difference between the top and bottom of the sea ice is very large. On the other hand, it is small in summer, when the temperature difference is small.

There is observational evidence that the magnitude of Arctic warming has varied with season. Chapman and Walsh (1993) examined the seasonal behavior of Arctic land station temperatures observed between 1961 and 1990. Their analysis shows that the median rates of warming in winter and spring are about 0.25°C and 0.5°C per decade, respectively. In summer the trend was near zero, and in autumn the trend was either neutral or showed slight cooling. The contrast between little or no observed warming in summer and substantial warming in winter and spring appears to be broadly consistent with the seasonal dependence of Arctic warming simulated by the model. Screen and Simonds (2010) used a state-of-the-art reanalysis product to estimate the seasonal variation of the trend of surface air temperature in high northern latitudes during the period 1989–2008. They found a pronounced near-surface warming trend over the Arctic Ocean in most seasons, but a much more modest near-surface warming in summer. They suggested that the large seasonal dependence of the warming was attributable mostly to the reduction of sea ice thickness.

Manabe et al. (2011) evaluated the seasonal dependence of the surface temperature trend during the past several decades, using a historical surface temperature data set compiled by the Climatic Research Unit of the University of East Anglia and the Hadley Centre for Climate Prediction and Research of the UK Meteorological Office. The result of their analysis is shown in figure 5.10, which illustrates the latitude/calendar-month distribution of the zonal mean surface air temperature anomaly for the period 1991–2009 relative to the 30-year base period of 1961–90. The anomalies over 1991–2009 can be regarded as indicative of the trend of zonal mean temperature during the past half-century. Zonal mean surface temperature increases at practically all latitudes and seasons. In the Northern Hemisphere, the warming trend increases with increasing latitude. The warming is particularly large over and around the Arctic Ocean during much of the year, with the notable exception of summer, when it is at its seasonal minimum. This is in good qualitative agreement with the model result shown in figure 5.8.

In sharp contrast with the Northern Hemisphere, polar amplification is not evident in the Southern Hemisphere. As a matter of fact, the temperature trend is small around 55° S in the Southern Ocean during much of the year, although the temperature anomalies are not shown poleward of 60° S in the figure because of the paucity of monthly mean temperature data in winter. The absence of significant warming in the Southern Ocean

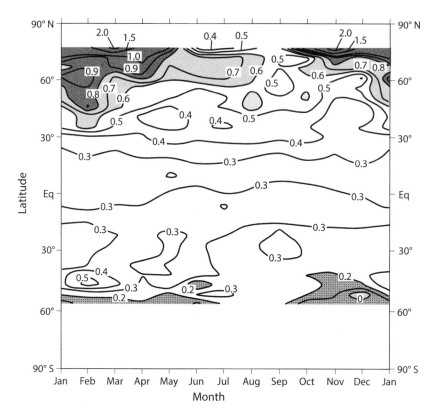

FIGURE 5.10 Observed zonal mean surface air temperature anomaly averaged over the period 1991–2009 as a function of latitude and calendar month. The anomaly represents the deviation from the average over the 30-year base period 1961–90. Because of poor data coverage, the anomaly is not shown poleward of 80° N or 60° S. From Manabe et al. (2011).

is attributable mainly to the exchange of heat between the mixed-layer and deep ocean, which cannot be captured by the atmosphere/mixed-layer model discussed here. We shall explore this subject further in chapter 8, using a coupled model in which a GCM of the atmosphere is combined with an ocean GCM.

Remote sensing of sea ice by microwave sounding instruments on satellites has made it possible to observe trends in areal coverage. Figure 5.11 illustrates the temporal variation of sea ice extent (%) in the Northern Hemisphere for September and March, when sea ice coverage is at a maximum and at a minimum, respectively (Perovich et al., 2010). Over the period 1979–2009, the extent of sea ice in the Arctic Ocean decreased at a rate of 8.9% per decade in September, whereas it decreased at much slower rate of 2.5% per decade in March. The observed difference in the rate of

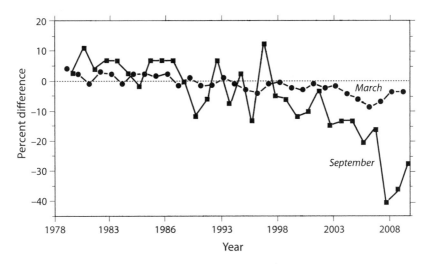

FIGURE 5.11 Time series of the difference (%) of observed sea ice extent from the mean value over the period 1979–2009 over the Arctic Ocean. Time series shown for September (the month of ice extent minimum) and March (the month of ice extent maximum). From Arndt et al. (2010).

change in sea ice coverage between September and March in the Arctic Ocean is consistent with the result obtained from the model. As shown in figure 5.9, the poleward retreat of the southern margin of sea ice in response to CO_2 quadrupling is much larger in September than in March.

The type of atmosphere/mixed-layer-ocean model described here was developed not only at GFDL but also at NASA/GISS by Hansen et al. (1983, 1984). The similarity of the latitude/calendar-month pattern of CO_2-induced warming from the GFDL model with that of the GISS model suggests that polar amplification is a robust result and not model dependent, at least in a qualitative sense. But although the basic structures of the GISS and GFDL atmosphere/mixed-layer models are similar, there are important differences between the two models.

The global mean surface temperature in the version of the GFDL model described above increased by about 4.1°C in response to the quadrupling of atmospheric CO_2. The log-linear relationship between CO_2 concentration and radiative forcing implies that the temperature change for CO_2 doubling would be slightly more than 2°C. In contrast, the model of Hansen et al. (1984) simulated a warming of 4°C for CO_2 doubling, meaning that it was almost twice as sensitive as the GFDL model presented here. The large difference between the sensitivities of the two models is indicative of the large uncertainty in the sensitivity of climate that still remains. The causes of this uncertainty will be the subject of discussion in the following chapter.

CHAPTER 6

Climate Sensitivity

One of the most challenging tasks of climate science is to obtain a reliable estimate of the sensitivity of climate, which is defined as the response of the global mean surface temperature to a specific thermal forcing, given a sufficiently long time (i.e., the equilibrium response). As pointed out in the preceding chapter, the sensitivities of the two early climate models constructed at GFDL and NASA/GISS were different from each other by a factor of about two. According to Flato et al. (2013), a similarly large difference still remains among the sensitivities of current climate models. The need to reduce this sizable uncertainty is one of the reasons why it is urgent to improve and validate the modeling of the radiative feedback processes that control the sensitivity of climate. In this chapter, we shall describe how the sensitivity of climate is determined through the interactions among various radiative feedback processes. We begin by obtaining a formula that relates the sensitivity of climate to the strength of the total radiative feedback that operates on the global-scale perturbation of surface temperature.

Radiative Feedback

The radiative heat balance of the Earth is maintained between the incoming solar radiation and outgoing radiation at the top of the atmosphere, as expressed by:

$$I = R, \tag{6.1}$$

where I denotes the global mean value of the incoming solar radiation, and R is that of outgoing radiation. The latter consists of two components, as expressed by:

$$R = L + S_r, \tag{6.2}$$

where L is the global mean value of outgoing longwave radiation, and S_r is that of reflected solar radiation at the top of the atmosphere. Suppose that a constant positive thermal forcing is applied continuously to the surface-atmosphere system that is coupled closely through the exchange of heat. Given a sufficiently long time, the temperature of this system will increase until the increase in the top-of-atmosphere (TOA) flux of outgoing radiation becomes equal to the thermal forcing, restoring the heat balance of the planet, as expressed by the following equation:

$$\Delta R = Q, \tag{6.3}$$

where ΔR is the change in the TOA flux of outgoing radiation and Q is the thermal forcing of the system. If Q is sufficiently small, the change in the global mean surface temperature, ΔT_s, would also be expected to be relatively small. In this case, ΔR is linearly proportional to ΔT_s, and may be expressed by:

$$\Delta R = \lambda \cdot \Delta T_s, \tag{6.4}$$

where λ is the feedback parameter, and denotes the global mean change of the TOA flux of outgoing radiation that accompanies a unit change in the global mean surface temperature, T_s. In other words, it represents the rate of radiative damping of the global-scale perturbation in surface temperature. It may be expressed by $\lambda = dR/dT_s$ or by substituting from equation (6.2) to yield:

$$\lambda = d(L + S_r)/dT_s. \tag{6.5}$$

Combining equations (6.3) and (6.4), one gets the following:

$$\Delta T_s = Q/\lambda. \tag{6.6}$$

This equation indicates that the sensitivity of climate is inversely proportional to the rate of damping of the global-scale perturbation of surface temperature. Simply stated, the larger the feedback parameter, the smaller the sensitivity of climate. The feedback parameter, as we have defined it, is a measure of the dependence of net outgoing radiation on global mean surface temperature. Thus, a positive value of λ indicates that the change in net outgoing radiation counteracts an imposed thermal forcing. Equation (6.6) yields the change in global mean surface temperature that would be required to restore the radiative balance of the planet. Using this equation, one can estimate the equilibrium response of global mean surface temperature to the radiative forcing.

Equations (6.5) and (6.6) were obtained by Wetherald and Manabe (1988) based on the assumption that the energy balance of the surface-atmosphere system is maintained between the net incoming solar radiation and the outgoing longwave radiation at the top of the atmosphere. It has been realized, however, that the stratosphere is essentially in a state of radiative equilibrium, with little heat exchange with the troposphere (i.e., across the tropopause), as in the 1-D radiative-convective model of the atmosphere introduced in chapter 3. For this reason, equation (6.6) is applicable not only to the surface-atmosphere system but also to the surface-troposphere system. Hereafter, we shall apply equation (6.6) to the latter, as the surface and the troposphere are coupled closely through heat exchange. Applying (6.6) in this way implicitly assumes that the rate of radiative damping of a perturbation in global mean surface temperature at the tropopause is equal to the rate of radiative damping at the top of the atmosphere.

It has become customary within the climate dynamics community to express the sensitivity of climate in terms of the equilibrium response of global mean temperature to the doubling of the atmospheric CO_2 concentration. Defined in this way, the equilibrium climate sensitivity $(\Delta_{2x} T_S)$ is given by:

$$\Delta_{2x} T_S = Q_{2x}/\lambda, \tag{6.7}$$

where Q_{2x} is the radiative forcing that results from a doubling of atmospheric CO_2. It should be noted here that the radiative forcing of the surface-troposphere system, Q_{2x}, differs from the instantaneous radiative forcing as determined at the top of the atmosphere. As shown in figure 3.5, stratospheric cooling occurs in response to CO_2 doubling, thereby reducing the outgoing longwave radiation from the top of the atmosphere. Because the stratosphere is essentially in radiative equilibrium, the net upward flux of radiation at the tropopause also decreases by the same amount, thereby reducing the radiative heat loss from the surface-troposphere system. For this reason, the radiative forcing of the surface-troposphere system is larger than it would be in the absence of stratospheric cooling, as was noted in chapter 3.

As equation (6.7) indicates, the sensitivity of climate is inversely proportional to the strength of the radiative damping that operates on the global-scale perturbation of surface temperature. To determine the sensitivity, it is therefore necessary to reliably estimate the strength of the radiative feedback that occurs in response to a perturbation of global mean surface temperature. This has been one of the most difficult problems to solve in climate modeling. Here, we shall attempt to estimate the strength of the radiative feedback and the sensitivity of climate, referring to the analysis of many global warming simulations along with other independent information.

The Gain Factor Metric

The most basic feedback in the climate system involves the change in the flux of outgoing radiation that would occur if the Earth emitted longwave radiation as a blackbody. In this idealized case, there would be a vertically uniform change of temperature, not only in the troposphere but also at the Earth's surface. This feedback yields a change in the TOA flux of outgoing longwave radiation that is in accordance with the Stefan-Boltzmann law of blackbody radiation, which states that total radiative output is proportional to the fourth power of the planetary emission temperature. This feedback is frequently termed the "Planck feedback."

A second kind of feedback involves the change in the TOA flux of total outgoing radiation due to other changes in the climate system that accompany a change in the global mean temperature at the Earth's surface. Examples of this type of feedback include changes in the absolute humidity, cloudiness, and the vertical profile of temperature in the troposphere, and the changes in snow cover and sea ice at the Earth's surface. These changes in turn affect the outgoing longwave radiation or the reflected solar radiation at the top of the atmosphere. Together, these two types of feedback determine the overall strength of the radiative feedback, thereby controlling the sensitivity of climate.

Based on this decomposition of feedback processes into two kinds, one can subdivide the feedback parameter (λ) into two components, as expressed by:

$$\lambda = \lambda_0 + \lambda_F, \tag{6.8}$$

where the first term, λ_0, indicates the strength of the basic Planck feedback defined above. It may be expressed as $\lambda_0 \approx 4\epsilon\sigma T_S^3$, where ϵ is the planetary emissivity, and σ is the Stefan-Boltzmann constant of blackbody radiation. The second term, λ_F, indicates the combined strength of the feedbacks of the second kind that involve the other changes listed in the preceding paragraph.

To characterize the relative contribution of the feedbacks of the second kind (λ_F) to the radiative feedback as a whole (λ), Hansen et al. (1984) introduced a nondimensional metric called the "gain factor," as defined by:

$$g = -\lambda_F / \lambda_0. \tag{6.9}$$

Using the gain factor, g, thus defined, the feedback parameter (λ) can be expressed as:

$$\lambda = \lambda_0 \cdot (1 - g). \tag{6.10}$$

As this equation indicates, the gain factor is a nondimensional metric that indicates the degree by which the basic Planck feedback is weakened by the feedbacks of the second kind, thereby enhancing the sensitivity of climate.

One should note the sign convention that results from this formalism. A positive value of the sensitivity parameter λ indicates that the net outgoing radiation increases in response to a surface warming. Thus, if a positive thermal forcing is imposed and the global mean surface temperature increases in response, the basic Planck feedback would partially counteract the effect of the forcing. The gain factor g is opposite in sign to λ, such that a positive value indicates a feedback mechanism that would enhance the effect of an imposed radiative forcing. The terms *positive feedback* and *negative feedback* that are commonly used in climate dynamics correspond to positive and negative values of the gain factor, respectively.

Inserting (6.10) into (6.7), one can obtain the following expression:

$$\Delta_{2x} T_S = Q_{2x} / [\lambda_0 \cdot (1 - g)]. \tag{6.11}$$

From equation (6.11), one can infer how the sensitivity of climate ($\Delta_{2x} T_S$) depends upon the gain factor. For example, if the gain factor is positive and is less than one, as is typically true in climate models, the feedback of the second kind partially compensates the basic Planck feedback and weakens the overall radiative damping, thereby enhancing the sensitivity of climate. As a matter of fact, the sensitivity increases nonlinearly at an accelerated pace as the gain factor increases linearly toward one. If the gain factor is equal to one, the feedback of the second kind compensates exactly with the basic Planck feedback, yielding a climate system with no radiative damping in operation. In this case, the global mean surface temperature would drift freely with no constraint. If the gain factor is larger than one, λ would be negative and equation (6.11) is no longer valid. In this case, climate would be unstable and a "runaway greenhouse effect" would be in operation. Realizing that climate has been stable over a very long time, it is not difficult to convince ourselves that the gain factor is less than one with radiative damping that is strong enough to keep our planet warm and habitable.

On the other hand, if the gain factor is zero, only the basic Planck feedback is in operation. In this case, the sensitivity is given by Q_{2x}/λ_0 and is relatively small. If the gain factor is negative, the feedback of the second kind enhances the Planck feedback, making the sensitivity of climate even smaller. To provide some perspective on an approximate range of values in which the actual gain factor may reside, we estimate, as examples, the gain factors of a few models that have been introduced in the preceding sections.

In the 1-D radiative-convective model introduced in chapter 3, the sensitivity of climate is 2.36°C when absolute humidity is allowed to respond to the change in temperature. In the case in which absolute humidity is fixed and no feedback involving water vapor is active, only the basic Planck feedback is in operation (i.e., $g = 0$), and the sensitivity is reduced to 1.33°C. From these two values, one can estimate the gain factor of the model from equation (6.11), which yields a gain factor of 0.44 for this case. This implies that the water vapor feedback in the model counteracts the basic Planck feedback, reducing the radiative damping by 44%. In other words, the strength of the overall feedback is reduced by a factor of 0.56 owing to the feedback of the second kind, leading to a sensitivity that is 1.77 (or 1/0.56) times as large as 1.33°C, the sensitivity of the model with only the basic feedback in operation.

A similar analysis can be made for the idealized GCM used in the CO_2-doubling experiment presented in chapter 5. In this model, the feedback of the second kind includes not only water vapor feedback, but also albedo and lapse rate feedback, though the latter is much smaller than the former in this model. The sensitivity of this model is 2.93°C and is substantially larger than the 2.36°C sensitivity of the radiative-convective model. Given that the sensitivity of this model is 1.33°C in the absence of these feedbacks with only the Planck feedback in operation (i.e., $g = 0$), one can estimate its gain factor, referring to equation (6.11). The gain factor thus obtained is 0.55 and is larger than 0.44 (i.e., the gain factor of the radiative-convective model). It is likely that the difference between the two gain factors is attributable mainly to the albedo feedback with a minor contribution from the lapse rate feedback in this model.

In both of these models, the combined effect of the feedbacks of the second kind counteracts the basic Planck feedback. Thus it reduces the strength of the overall feedback, increasing substantially the sensitivity of model climate. In the following section, we shall identify various feedback processes of the second kind that interact with one another, thereby affecting the sensitivity of climate. First, we formulate the relationship between the gain factors of the individual feedbacks of the second kind and the sensitivity of climate.

As given by equation (6.8), the feedback parameter, λ, can be expressed as the sum of the basic Planck feedback, λ_0, and λ_F, which can be expressed as the sum of contributions from various feedbacks of the second kind, as in Wetherald and Manabe (1988):

$$\lambda_F = \lambda_\Gamma + \lambda_w + \lambda_c + \lambda_a, \tag{6.12}$$

where $\lambda_\Gamma, \lambda_w, \lambda_c,$ and λ_a denote, respectively, the changes in the TOA flux of outgoing radiation due to changes in the vertical lapse rate of temperature (Γ), water vapor (w), cloud (c) in the troposphere, and albedo (a) at the Earth's surface that would accompany a unit (i.e., 1°C) change in the global mean surface temperature. Although there are other feedbacks of the second kind, they are not included here because their contributions are relatively small. Dividing both sides of the equation by $-\lambda_0$—that is, the strength of the basic Planck feedback—one can express the overall gain factor as the sum of the gain factors associated with each feedback of the second kind:

$$g = g_\Gamma + g_w + g_c + g_a, \tag{6.13}$$

where:

$$\begin{pmatrix} g_\Gamma \\ g_w \\ g_c \\ g_a \end{pmatrix} = -\frac{1}{\lambda_0} \begin{pmatrix} \lambda_\Gamma \\ \lambda_w \\ \lambda_c \\ \lambda_a \end{pmatrix}. \tag{6.14}$$

Inserting equation (6.13) into (6.11), one gets the following equation, which indicates the relationship between the sensitivity of climate, $\Delta_{2x}T_S$, and the gain factors of the individual feedback processes, such as lapse rate feedback, water vapor feedback, cloud feedback, and albedo feedback:

$$\Delta_{2x}T_S = Q_{2x}/\{\lambda_0 \cdot [1 - (g_\Gamma + g_w + g_c + g_a)]\}. \tag{6.15}$$

Feedbacks of the Second Kind

Using the formula obtained above, we shall describe here how various feedbacks of the second kind interact with one another, thereby controlling the sensitivity of climate.

Lapse Rate Feedback

When temperature increases at the Earth's surface in response to positive thermal forcing, it also increases in the troposphere, where heat is transported vertically through deep convection and large-scale circulation. The magnitude of the warming usually varies with height, thereby changing the vertical lapse rate of temperature. In the 3-D model introduced in chapter 5, for example, the overall global warming is accompanied by a decrease in the vertical lapse rate in low latitudes, where deep moist convection predominates, and an increase in lapse rate at high latitudes, where warming is strongest near the surface. Thus, the lapse rate hardly changes when averaged over the entire model domain.

Consider a case in which the response to a radiative forcing is a warming of the troposphere that increases with height (i.e., a decreased lapse rate), as in the low-latitude response of the model discussed above. The increase in the TOA flux of outgoing longwave radiation in this case would be larger than in a case with the same surface warming and a vertically uniform temperature change. Thus, a decrease in lapse rate increases the strength of the radiative damping defined by equation (6.5) and the sensitivity of climate decreases as expressed by equation (6.7). Conversely, if the thermal response is a warming that decreases with height (i.e., an increased lapse rate), the increase in TOA flux of outgoing longwave radiation would be smaller than it would be in the case in which the warming throughout the troposphere is the same as the surface warming. Thus, an increase in lapse rate weakens the radiative damping and increases the climate sensitivity. Hereafter, the feedback process that involves a change in the vertical lapse rate of temperature in the troposphere is called "lapse rate feedback."

Water Vapor Feedback

Water vapor has short residence time of a few weeks in the troposphere as it evaporates from the Earth's surface and is removed through precipitation. When an air parcel moves upward in the atmosphere, pressure decreases with increasing height and the parcel cools owing to adiabatic expansion. Eventually, water vapor in the parcel condenses, yielding precipitation. On the other hand, a downward-moving parcel warms owing to adiabatic compression, reducing its relative humidity. In short, the distribution of relative humidity in the troposphere is largely controlled by the vertical movement of air. As long as the magnitude of global warming is modest, it is likely that changes in the general circulation of the atmosphere would

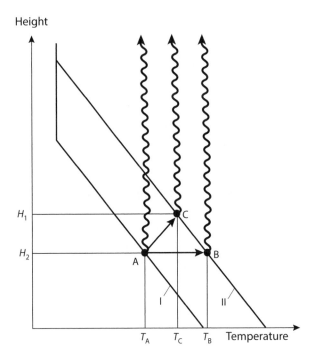

FIGURE 6.1 Diagram illustrating schematically how water vapor feedback shifts the effective center of planetary emission of longwave radiation upward, thereby enhancing the magnitude of global warming. Slanted lines I and II indicate the vertical temperature profiles in the troposphere before and after the warming, respectively. (The vertical line segment represents the vertical temperature profile of the almost isothermal lower stratosphere.) Black dots on the slanted lines indicate the effective planetary emission centers of the outgoing longwave radiation from the top of the atmosphere. See main text for further explanation.

be small, with little change in distribution of relative humidity (e.g., Held and Soden, 2000). With generally little change in relative humidity, the absolute humidity of air would increase with increasing temperature in the troposphere, thereby enhancing its infrared opacity.

Increasing the infrared opacity of the troposphere affects the outgoing longwave radiation at the top of the atmosphere. The effective emission center of the TOA flux of outgoing longwave radiation would increase in altitude, as explained in chapter 1 in the discussion of the effect of greenhouse gases on longwave radiation. Because temperature decreases with increasing height in the troposphere, the increase in altitude of the effective emission center results in a reduction of the outgoing longwave radiation from the top of the atmosphere. This response is illustrated in figure 6.1. The thick solid line I in this figure indicates schematically the vertical profile of temperature that decreases almost linearly with increasing height in the troposphere and is almost isothermal in the overlying lower strato-sphere. The black dot (A) placed on line I indicates the effective emission center of the upward flux of outgoing longwave radiation that reaches the top of the atmosphere. Suppose that the temperature of the troposphere increases from I to II without changing its lapse rate. If water vapor feed-back is absent, the absolute humidity does not change, thereby keeping the infrared opacity of the troposphere unchanged. In this case, the effective emission center of outgoing longwave radiation (B) would remain at the same altitude, H_2, and its temperature increases from T_A to T_B. If water vapor feedback is in operation, on the other hand, increasing the infrared opacity of the troposphere, the effective emission center of outgoing long-wave radiation (C) would rise from H_2 to H_1 and its temperature would increase from T_A to T_C, as indicated schematically in the figure.

In this manner, water vapor feedback reduces the magnitude of a change in the emission temperature in response to a given change in surface tem-perature. In other words, water vapor feedback reduces the strength of the longwave feedback that operates on the perturbation of surface tempera-ture, thereby enhancing the sensitivity of climate, as defined by equation (6.7). For further discussion of the role of the water vapor feedback in the sensitivity and variability of climate, see the study conducted by Hall and Manabe (1999).

In addition to absorbing and emitting longwave radiation as noted above, water vapor absorbs incoming and reflected solar radiation in the near-infrared portion of the solar spectrum between 0.8 and 4 μm (fig-ure 1.6d). If the temperature of the surface-troposphere system increases, it is expected that absolute humidity would increase as explained above, enhancing the absorption of solar radiation. For this reason, the upward TOA flux of reflected solar radiation decreases with increasing surface temperature. In short, the shortwave component of water vapor feedback also helps to reduce the overall strength of the radiative feedback that op-erates on the perturbation of surface temperature, thereby enhancing the sensitivity of climate. According to the study conducted by Wetherald and Manabe (1988), for example, the magnitude of the enhancement is about one-fifth of the longwave component and is relatively small.

Albedo Feedback

Snow and sea ice reflect a large fraction of the incoming solar radiation that reaches the Earth's surface. When temperature increases at the Earth's surface, the area covered by snow and sea ice usually decreases, reducing the global surface albedo. Thus, the TOA flux of reflected solar radiation usually decreases with increasing surface temperature, weakening the overall strength of the radiative damping that operates on the perturbation of the global mean surface temperature. In short, the gain factor associated with the albedo feedback of snow and sea ice is positive, enhancing the sensitivity of climate.

The albedo feedback of snow and sea ice operates on a relatively short time scale. Over longer time periods, snow cover that survives from one year to the next has the potential to build up and eventually transform into a continental ice sheet. A model that incorporates the albedo feedback of continental ice sheets as well as snow and sea ice is important for studying the transitions between glacial and interglacial periods that have dominated the climate record during the past several million years. Initial attempts to do this were made by Weertman (1964, 1976), Pollard (1978, 1984), Berger et al. (1990), and Deblonde and Peltier (1991). Although these studies used relatively simple EBMs of climate, they have been very successful for exploring the role of ice sheet albedo feedback in glacial-interglacial climate transitions. Aided by the remarkable advancement of computer technology, it is becoming possible to study the glacial-interglacial transition of climate using GCMs of the coupled atmosphere-ocean-cryosphere system that incorporate explicitly the dynamics and thermodynamics of continental ice sheets (e.g., Gregory et al., 2012).

Cloud Feedback

Clouds consist of innumerous water droplets or ice crystals of various sizes. When light passes from air to water or from water back to air, it changes direction as it crosses the interface between the two media. This process is called refraction. As light travels through a water droplet, it will be refracted twice, once as it enters the droplet and once as it exits from the droplet. This is the process by which light is scattered by cloud droplets; repeated encounters with cloud droplets can completely turn the light around. Thus, clouds reflect a substantial fraction of incoming solar radiation, exerting a cooling effect upon the heat budget of the planet. Because an ice surface also has large reflectivity, ice clouds have a qualitatively similar effect.

Although water is almost transparent to the visible portion of solar radiation, it absorbs longwave radiation very strongly, as does an ice surface. This is the main reason why most cloud (with exception of thin cloud) absorbs almost completely longwave radiation and emits it almost as a blackbody would according to Kirchhoff's law. For example, clouds absorb almost all of the upward flux of longwave radiation from below, whereas they emit upward flux almost as a blackbody. Because the temperature of the subcloud layer and that of the Earth's surface are usually higher than that of the cloud top, the upward flux of longwave radiation at the cloud bottom is usually larger than the upward flux at the top. For this reason, the TOA flux of outgoing longwave radiation is usually smaller in the presence of cloud than in its absence. In short, clouds trap a substantial fraction of the upward flux of longwave radiation emitted by the Earth's surface before it reaches the top of the atmosphere, exerting a greenhouse effect upon the heat budget of the planet.

Using satellite measurements of the TOA flux of outgoing longwave radiation and that of reflected solar radiation over all-sky and clear-sky conditions, Harrison et al. (1990) estimated the effect of clouds upon the radiation budget of the Earth. According to their estimate, heat loss due to the reflection of incoming solar radiation by cloud is about 48 W m^{-2}, which is about 47% of the 102 W m^{-2} that is the total reflected solar radiation at the top of the atmosphere. On the other hand, the heat gain due to the greenhouse effect of cloud is about 31 W m^{-2}, which is about 20% of the 151 W m^{-2} that is the total greenhouse effect of the atmosphere. The net heat loss from these two opposing effects is 17 W m^{-2}, which is more than four times as large as the ~4 W m^{-2} that is the thermal forcing that results from the doubling of the CO_2 concentration in the atmosphere. This implies that a 25% increase of total cloud amount would be large enough to compensate for the warming due to CO_2 doubling, if cloud properties did not otherwise change. It also implies that a reduction of total cloud amount by 25% could increase the magnitude of the warming by a factor of about two. (See Ramanathan and Coakley [1978] for further discussion of this subject.)

The outgoing radiation at the top of the atmosphere is likely to depend not only on the distribution of clouds but also on their microphysical properties, such as the size and number density of cloud droplets. Somerville and Remer (1984) speculated that the microphysical properties of clouds might change as temperature increases in the atmosphere owing to global warming. Referring to the study of Feigelson (1978), which was based upon 20,000 measurements made in the former Soviet Union, they suggested that the liquid water content of stratus clouds is likely to increase as the saturation vapor pressure of air increases with increasing temperature. Because clouds of higher liquid water content would be optically thicker for the same depth, they would reflect more solar radiation. Thus, the TOA flux of reflected solar radiation would be likely to increase with increasing temperature, enhancing the radiative damping of surface temperature perturbation and thereby reducing the sensitivity of climate.

More recent observational studies have found that low cloud in the tropics reflects less solar radiation as temperature increases, although this is partially offset by the response of such cloud to changes in the strength of the temperature inversion (Klein et al., 2017). Thus, there is more uncertainty about both the sign and magnitude of cloud feedback in the real climate system. Nevertheless, these studies underscore the possibility that the microphysical properties of clouds may change systematically as temperature increases owing to global warming, thereby affecting the strength of radiative feedback and, accordingly, the sensitivity of climate.

Feedback in 3-D Models

This section presents a quantitative analysis of the radiative feedback processes that operate in some 3-D climate models that have been constructed. We begin with the pioneering study of radiative feedback conducted by Hansen et al. (1984), using the atmosphere/mixed-layer-ocean model constructed at GISS. Introducing the nondimensional gain factor described earlier in this chapter, they evaluated quantitatively the relative contributions of various radiative feedback processes to the sensitivity of climate.

The model used for their study was constructed by combining a GCM of the atmosphere with a heat balance model of the mixed-layer ocean and of the continental surface. This model is generally similar to that developed at GFDL by Manabe and Stouffer (1979, 1980) for the CO_2-quadrupling experiment described in chapter 5. The two models, however, differ from each other in a few important respects. Although the distribution of cloud is held fixed in the early version of the GFDL model, it is predicted in the GISS model, which means that cloud feedback is in operation. Another very important difference is the so-called "Q-flux" technique employed by Hansen et al. (1983, 1984). In the GFDL model, it is assumed that heat exchange between the mixed layer and the deep subsurface layer of the ocean is absent. In the GISS model, on the other hand, a time-invariant heat flux is prescribed and applied to the mixed layer, such that the geographic distribution of sea surface temperature is realistic in the control run. An identical heat flux is also prescribed for the doubled-CO_2 run, implicitly assuming that ocean heat transport as well as other systematic biases of the model remain unchanged despite the CO_2 doubling. Because the strength of the albedo feedback of snow and sea ice depends critically upon the distribution of temperature at the Earth's surface, the GISS model is well suited for the study of climate sensitivity.

The numerical time integration of the GISS model described here was performed for two different CO_2 concentrations (315 and 630 ppmv) over a sufficiently long time to achieve quasi-equilibrium. From the difference between the two states thus obtained, Hansen et al. (1984) estimated the equilibrium response of temperature to the doubling of the atmospheric CO_2 concentration. They found that the global mean temperature at the Earth's surface increased by 4.2°C in response to CO_2 doubling. Assuming that the thermal forcing of the troposphere-surface system (Q_{2x}) is ~4 W m^{-2} and the strength of the basic feedback (λ_0) is 3.21 W m^{-2}, one can estimate a gain factor of 0.70 for this model using equation (6.11). This implies that the combined effect of the feedbacks of the second kind counteracts the basic Planck feedback, reducing the strength of the overall radiative damping by as much as 70%, with 30% remaining. In other words, the sensitivity of the GISS model is 3.3 (= 1/0.3) times as large as the sensitivity of a model with only the basic Planck feedback in operation.

The overall gain factor may be expressed as the sum of gain factors of feedbacks of the second kind as expressed by equation (6.13). Using the output from their CO_2-doubling experiment, they extracted the global mean changes of absolute humidity, cloudiness, vertical lapse rate of temperature in the troposphere, and albedo of the Earth's surface. Inserting these changes, one by one or in combination, into a 1-D model of radiative-convective equilibrium, they obtained changes in global mean temperature. From the changes thus obtained, they estimated, albeit roughly, the gain factors of the various feedbacks of the second kind, which are listed in the second column of table 6.1.

The gain factors of the water vapor feedback and the lapse rate feedback are presented both separately and in combination ($g_w + g_r$). The combined feedback is presented because a change in the vertical gradient of temperature usually induces a change in that of absolute humidity such that the change in the TOA flux of longwave radiation due to the former partially offsets the change due to the latter. Because of the partial compensation between the two feedbacks, the gain factor of the combined water vapor–lapse rate feedback is substantially smaller than that of the water vapor feedback alone. In view of the close interaction between the two feedbacks in their model, Hansen et al. combined them into a single category called "combined water vapor–lapse rate feedback."

The gain factor of the combined water vapor–lapse rate feedback of the GISS model is 0.40 and is larger than those of the cloud feedback and albedo feedback. It is smaller, but comparable in magnitude to the 0.44 gain factor of the water vapor feedback in the radiative-convective model introduced in chapter 3, in which other feedbacks of the second kind are absent. The gain factor of the cloud feedback is 0.22. Although it is smaller than the combined water vapor–lapse rate feedback, it is more than twice as

TABLE 6.1 *Average total gain factor and gain factors*
of individual feedbacks of the second kind

	Hansen et al. (1984)	Colman (2003)	Soden and Held (2006)
g_w	0.57	0.53 ± 0.12	0.56 ± 0.06
g_Γ	−0.17	-0.10 ± 0.12	-0.26 ± 0.08
$g_w + g_\Gamma$	0.40	0.43 ± 0.06	0.30 ± 0.03
g_c	0.22	0.17 ± 0.10	0.21 ± 0.11
g_a	0.09	0.09 ± 0.04	0.08 ± 0.02
g	0.71	0.69 ± 0.08	0.59 ± 0.12

The standard deviation (with ± sign) is added to the average gain factors as an indicator of the spread among the climate models.

g, gain factor of total feedback—i.e., the sum of the gain factors of the feedbacks of the second kind; g_w, gain factor of water vapor feedback; g_Γ, gain factor of lapse-rate feedback; $g_w + g_\Gamma$, gain factor of combined water vapor–lapse-rate feedback; g_c, gain factor of cloud feedback; g_a gain factor of albedo feedback.

Each of the gain factors listed here is calculated using equation (6.14). The value of λ_0 used for the computation is 3.21 W m^{-2}; i.e., the average value of the AR4 models (the 19 models used for the IPCC *Fourth Assessment Report*, and for the analysis presented here).

large as the 0.09 gain factor for the albedo feedback. Adding these gain factors yields a total of 0.71 for the feedbacks of the second kind. Inserting this value into equation (6.15) results in an estimate of 4.3°C as the sensitivity of the GISS model, which is very close to the actual sensitivity, suggesting that the use of the 1-D model is a reasonable approach to estimating the feedbacks of the second kind.

Figure 6.2 illustrates how the sensitivity of climate increases at an accelerated pace, according to equation (6.15), as the gain factors of each feedback of the second kind are added one by one. For example, if only the basic Planck feedback were in operation, the sensitivity would be 1.25°C. With the addition of the combined water vapor–lapse rate feedback, it increases from 1.25°C to 2.1°C as the gain factor increases from 0 to 0.40. With the addition of the albedo feedback, sensitivity increases from 2.1°C to 2.45°C as the overall gain factor increases from 0.4 to 0.49. Finally, with the addition of the cloud feedback, the sensitivity increases from 2.45°C to 4.3°C as the gain factor increases from 0.49 to 0.71.

One should note here that the sensitivity of climate increases nonlinearly as the gain factor in the denominator in equation (6.11) approaches one, reducing the strength of the overall radiative damping. Thus, the contribution of each feedback of the second kind to the sensitivity depends not only on the sign and the magnitude of its gain factor, but also on the order in which it is added. Nevertheless, the result presented here indicates that the

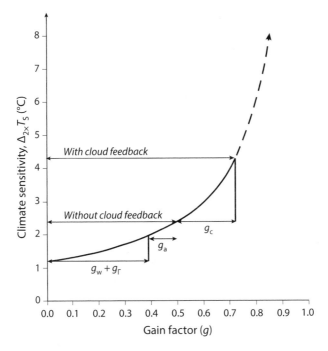

FIGURE 6.2 Relationship between gain factor (g) and the sensitivity of climate ($\Delta_{2x}T_S$) as expressed by equation (6.15). Climate sensitivity increases at an accelerated pace as each gain factor is added. See main text for further explanation.

sensitivity of the GISS model with cloud feedback (4.3°C) is substantially larger than that of the model without it (2.45°C).

Analyzing the results from their CO_2-doubling experiments, Hansen et al. found that the positive feedback effect of clouds is attributable to two aspects of their response to global warming. The first aspect is a reduction in total cloud amount. As explained earlier, cloud has two opposing effects. On the one hand, it reflects incoming solar radiation, thereby increasing the TOA flux of outgoing reflected solar radiation. On the other hand, it exerts a greenhouse effect, reducing the TOA flux of outgoing longwave radiation. Since the former effect is usually larger than the latter, a reduction of total cloud amount usually results in a reduction in the TOA flux of total outgoing radiation.

The second aspect of the response of cloud that contributes to a positive feedback effect is an increase in cloud-top height. Because temperature decreases with increasing height in the troposphere, an increase in cloud-top height also reduces the TOA flux of outgoing radiation, as longwave radiation from the cloud tops is being emitted at a lower temperature. In summary, the TOA flux of total outgoing radiation decreases not only because of the reduction in total cloud amount but also because of the increase in cloud-top height. This explains why the cloud feedback of the GISS model is positive, thereby enhancing the sensitivity of climate.

At about the same time as the GISS study described above, Wetherald and Manabe (1980, 1986, 1988) conducted modeling studies of cloud feedback using two different models developed at GFDL. To understand the similarities and differences between the response of clouds in the models developed at these two climate modeling centers, we present an analysis of cloud feedback in one of the GFDL models (Wetherald and Manabe, 1988) and compare it with the GISS model.

The GFDL model used for this study was constructed by modifying the atmosphere/mixed-layer-ocean model of Manabe and Stouffer (1979, 1980) that was described in chapter 5. In contrast to the original version of the model, in which the distribution of cloud was prescribed and held fixed throughout the course of a time integration, cloud cover was predicted in the modified version. Overcast cloud was placed at each grid point where relative humidity exceeded a specified critical percentage. Using this model, Wetherald and Manabe obtained two quasi-equilibrium states for the standard and doubled concentrations of atmospheric CO_2. From the difference between the two states, they determined how the distribution of cloud changes in response to a doubling of atmospheric CO_2 concentration, thereby altering the upward flux of outgoing longwave radiation and reflected solar radiation from the top of the atmosphere.

Figure 6.3a illustrates the distribution of annually averaged zonal mean cloud fraction obtained from the control experiment with a standard concentration of atmospheric CO_2. The simulated cloud fraction is very small in the stratosphere above the tropopause, which is indicated by the thick dashed line. On the other hand, the cloud fraction is relatively large in the upper troposphere, where high cloud is usually observed in the actual atmosphere. It is also relatively large in a thin layer several hundred meters above the Earth's surface, where stratus clouds often appear. Between these layers of relatively high cloud fraction in the upper and lower troposphere, there is a thick layer in the lower-middle troposphere around 700 hPa in which the cloud fraction is small.

Upon close inspection, one can see that the fraction of high cloud in the upper troposphere is at a maximum in the tropics near the ITCZ and in the zonal belt between 40° and 70° latitude, where midlatitude storm tracks are located. In these regions, humid air is often carried upward until it reaches the upper troposphere, where it spreads horizontally, maintaining a thick layer of high cloud. In contrast, the cloud fraction is relatively low in the

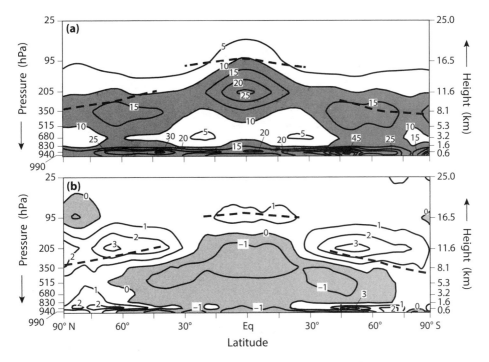

FIGURE 6.3 Latitude-height (pressure) profiles of (*a*) annually averaged zonal mean cloud fraction (%) obtained from the control simulation, and (*b*) the change in annually averaged zonal mean cloud fraction (%) from the control run (1×CO$_2$) to the CO$_2$-doubling run (2×CO$_2$). Thick dashed lines indicate the approximate height of the tropopause. The approximate heights (km) of the finite difference levels of the model are indicated on the right-hand side of the figure. From Wetherald and Manabe (1988).

middle troposphere, where the variance of vertical velocity is large and zonal mean relative humidity is reduced by the adiabatic compression of sinking air. Relative humidity and cloud fraction are high in the planetary boundary layer immediately adjacent to the surface, where water vapor is replenished continuously through evaporation from the underlying surface and air often cools owing to moist convective adjustment (see chapter 4) that transfers heat from the boundary layer to the middle and upper troposphere of the model.

It is quite encouraging that the annual cloud fraction described here resembles the corresponding profile of cloud occurrence probability obtained from the recent NASA Cloud-Aerosol Lidar and Infrared Pathfinder Satellite Observation (CALIPSO) mission, shown in figure 6.4, despite the simplicity of the parameterizations used in the model. In this version of the GFDL model, cloud is placed in those grid points where relative humidity exceeds a certain critical value (i.e., 99%), otherwise, a grid point

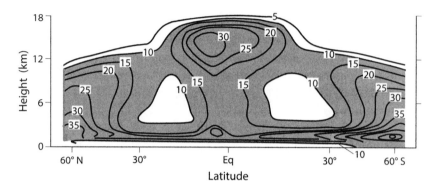

FIGURE 6.4 Latitude-height profile of annual zonal mean cloud occurrence (%) obtained from the recent NASA satellite observation mission CALIPSO. From Boucher et al. (2013).

is cloud-free. In other words, the 3-D distribution of cloud is based solely upon the relative humidity at each grid point. Realizing that this cloud prediction scheme is very simple, it is quite surprising that the simulated distributions of both high and low cloud resemble those obtained from CALIPSO. The resemblance suggests that the distribution of cloud is largely controlled by the large-scale circulation in the atmosphere and does not depend critically on the microphysical details of formation, maintenance, and disappearance of cloud cover.

Figure 6.3b shows the simulated change in zonal mean cloud fraction that occurs in response to the doubling of the CO_2 concentration in the atmosphere. Comparing it with figure 6.3a, cloud fraction increases in the upper half of the layer of high cloud in the control experiment, whereas it decreases in the lower half, implying that high cloud shifts upward in low and middle latitudes. This upward shift is also evident in figure 6.5, which illustrates the vertical profiles of zonal mean cloud fraction at 31° N obtained from the $1\times CO_2$ and $2\times CO_2$ runs. As noted earlier in this chapter, an upward shift of high cloud reduces the upward TOA flux of longwave radiation. Thus it helps to weaken the overall radiative damping, thereby enhancing the sensitivity of climate.

The change in zonal mean cloud fraction shown in figure 6.3b also features a decrease in cloud fraction at low latitudes in much of the free atmosphere, whereas it tends to increase in middle and high latitudes. In the planetary boundary layer, on the other hand, low cloud increases at most latitudes, particularly in the middle and high latitudes. In response to these changes, the TOA flux of outgoing reflected solar radiation decreases equatorward of 40° latitude, whereas it increases poleward of this latitude in both hemispheres. Since the former area is substantially larger than the

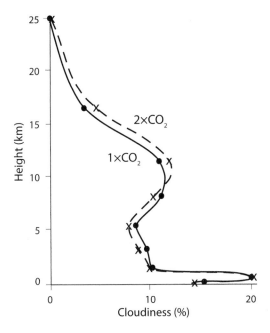

FIGURE 6.5 Vertical distribution of zonal mean cloud fraction at 30° N obtained from the control (1×CO$_2$) and CO$_2$-doubling (2×CO$_2$) experiments.

latter, the global mean TOA flux of reflected solar radiation decreases in response to the change in the zonal mean cloudiness shown in figure 6.3b. Thus, it also weakens the overall radiative damping, thereby enhancing the sensitivity of climate.

As described above, the simulated distribution of cloud changes systematically, reducing not only the global mean TOA flux of outgoing longwave radiation but also that of reflected solar radiation. Thus, the cloud feedback would be expected to have a positive gain factor, enhancing the sensitivity of climate. The gain factors of the cloud feedback in the GFDL model were estimated from the changes in the upward TOA fluxes of longwave radiation and reflected shortwave radiation that result from the change in the amount and distribution of cloud. The longwave and shortwave gain factors of cloud feedback thus obtained have small positive values of 0.04 and 0.08, respectively. Adding these two gain factors yields a value of 0.12 for the total gain factor of the cloud feedback in the GFDL model. Although this gain factor is positive, in qualitative agreement with the GISS model, its magnitude is only about half as large as the gain factor of 0.22 obtained from the GISS model. If the GISS model had the same gain factor of cloud feedback as the GFDL model, its gain factor of total feedback would have been 0.60 and its sensitivity would be 3.1°C. This is substantially smaller than its actual sensitivity of 4.2°C. Therefore, it is worthwhile inquiring why the strength of the cloud feedback is so different between these two models.

Comparing the changes in zonal mean cloudiness obtained from the GISS and GFDL models, one can identify many common features. For example, cloud-top height increases in both models. Cloud amount in the free atmosphere decreases in low and middle latitudes in both models, whereas it increases in high latitudes. However, the reduction of cloud amount is larger in the GISS model than in the GFDL model. Of particular interest is the increase in low cloud amount in the middle and high latitudes in the GFDL model, which is particularly large in the Southern Hemisphere. Similar changes in low cloudiness are not evident in the GISS model, though overall cloud amount increases in the lower troposphere in high latitudes of both hemispheres. Averaged globally, cloud amount decreases in the GISS model, whereas it hardly changes in the GFDL model. This is likely to be an important reason why the gain factor for cloud feedback in the GISS model is much larger than in the GFDL model.

In both the GISS and GFDL models described here, the cloud feedback involves a change in the distribution of cloud that depends solely upon the distribution of relative humidity. In the simple parameterizations used in both models, cloud feedback does not involve changes in cloud optical properties. Despite this similarity in cloud treatment, the gain factor of the cloud feedback in the GISS model is almost twice as large as in the GFDL model. This suggests that the large intermodel difference in the strength of the cloud feedback of the current climate models is attributable in no small part to the change in the distribution of cloud that accompanies global warming. For example, Soden and Vecchi (2011) found that the variations in the strength of cloud feedback among models were due primarily to changes in low cloud amount, rather than low cloud optical properties.

The strength of the radiative feedbacks of more recent climate models has been the subject of comprehensive analysis by Colman (2003) and Soden and Held (2006). Their analyses have been very useful in assessing the uncertainties in the strength of such feedbacks and the sensitivity of climate. Colman estimated the average feedback parameters for the 10 models constructed prior to the end of the twentieth century. These models (which we will call the "early models") include not only the GISS and GFDL models described above but also other models in which the optical properties of clouds are determined by cloud microphysical parameterization. Using equation (6.14), one can convert the feedback parameters obtained by Colman into gain factors. The average gain factors thus obtained, along with the gain factors of Hansen et al. (1984), are listed in table 6.1. Soden and Held (2006) also estimated feedback parameters for the 19 models used for the IPCC *Fourth Assessment Report* (hereafter the "AR4 models"). These feedback parameters were converted into gain factors and are also listed in table 6.1. Referring to the table, we shall review the gain factors of many climate models that have been constructed during the past several decades.

The average gain factor of the combined water vapor–lapse rate feedback of the early models is 0.43 and that of the AR4 models is 0.30. The difference between the two gain factors is attributable mainly to the difference in the average gain factor of the lapse rate feedback. As table 6.1 indicates, the average gain factor of the water vapor feedback of the early models is 0.53, which is similar to the corresponding gain factor of 0.56 for the AR4 models. However, the average gain factor of the lapse rate feedback of the early models is −0.10 and differs by a factor of more than two from the gain factor of −0.26 for the AR4 models. This large difference in the average lapse rate feedback between the two sets of models implies that the reduction in vertical temperature gradient per unit increase in global mean surface temperature is less in the early models than in a majority of the AR4 models.

The recent trend of the vertical lapse rate in the troposphere has been the subject of observational studies by Fu et al. (2011) and Po-Chedley and Fu (2012). Using retrievals of atmospheric temperature from microwave sounding units on a series of satellites, they estimated the trend of the vertical temperature gradient in the tropics. Their analyses indicate that most of the AR4 models overestimate substantially the increase in the static stability of the upper troposphere that has occurred in low latitudes during the past several decades. This implies that the gain factors of the lapse rate feedback in the majority of the AR4 models may be too large by a substantial factor. Although the gain factor of the lapse rate feedback partially compensates for that of the water vapor feedback as noted previously, the average gain factor of the combined water vapor–lapse rate feedback may be substantially larger than 0.3, which is the average value of the AR4 models.

One of the most important factors that control the vertical temperature gradient is deep moist convection. The large difference between the average gain factors of the lapse rate feedback between the two sets of models implies that the parameterization of deep moist convection differs significantly between them. Moist convection affects not only the vertical distribution of temperature but also those of relative humidity and cloudiness. Thus, it affects substantially the strength of water vapor feedback and cloud feedback, as well as that of the lapse rate feedback. A major effort needs to be devoted to the improvement and validation of the parameterization of moist convection for the reliable determination of climate sensitivity.

The average gain factor of cloud feedback for the early models is 0.17 ± 0.10 and is similar to the corresponding gain factor (0.21 ± 0.11) of the AR4 models. The standard deviations attached to these gain factors are large enough to include those obtained from the GISS (0.22) and GFDL (0.11) models. The relatively large standard deviations attached to the gain factors of cloud feedback indicate that a large spread exists among models. Conducting an intercomparison of the modeled feedback a few decades ago, Cess et al. (1990) reached a similar conclusion. In order to reduce the large uncertainty in the gain factor of the cloud feedback, one can validate the modeled cloud feedback using satellite observations of the outgoing radiation from the top of the atmosphere.

The average gain factor of the albedo feedback is relatively small and more consistent across the models examined. The average gain factor of the early models obtained by Colman is 0.09 ± 0.04, in good agreement with the 0.09 obtained by Hansen et al. (1984). Similar average gain factors of albedo feedback of 0.08 ± 0.02 and 0.09 ± 0.03 were obtained from slightly different subsets of the AR4 models by Soden and Held (2006) and Winton (2006), respectively.

Summing the average gain factors of the individual feedbacks, one can get an average gain factor of the overall radiative feedback. The average gain factor of the early models is 0.69 ± 0.08 and that of the AR4 models is 0.59 ± 0.12. From these values, one can estimate the sensitivity for these two sets of models using equation (6.11). The sensitivity obtained from the average values is 3.8°C for the early models and 3.0°C for the AR4 models. As implied by the large standard deviations of the overall gain factors, the sensitivity varies greatly in both sets of models. If the gain factors were normally distributed, this would imply that two-thirds of the early and AR4 models have sensitivities from 3.2°C–5.4°C and 2.4°C–4.3°C, respectively, with the remainder lying outside these ranges.

In view of the large spread among the sensitivities of climate models, it is difficult to estimate the sensitivity of climate based upon the results of the numerical experiments alone. For this reason, it is desirable to estimate the sensitivity using other, independent information. For example, Wigley et al. (2005) estimated the sensitivity of climate from the temporal variation of the global mean surface temperature observed immediately following large volcanic eruptions. As we know, volcanic eruptions can release large amounts of sulfur dioxide that are transformed into sulfate aerosols, exerting a temporary global mean cooling effect at the surface. Using simple EBMs with wide-ranging sensitivities, they performed a set of numerical experiments and found that the best simulation of the temporary volcanic cooling was achieved with a model that had a sensitivity of ~3°C.

Another promising approach for estimating the sensitivity of climate is to use data from the geologic past. Many attempts have been made to simulate glacial-interglacial differences of sea surface temperature, using a climate model with known sensitivity. Comparing the simulated differences with the actual differences reconstructed from paleoclimatic data, one can estimate the sensitivity of climate. We will present examples of such studies in the next chapter.

In the future, it is highly desirable to continue the satellite observations of the TOA fluxes of outgoing longwave radiation and reflected solar radiation, not only over all sky but also over clear sky (Barkstrom, 1984; Loeb et al., 2009; Wielicki et al., 1996) and over a span of many decades. Using long-term observations of the global mean TOA fluxes and global mean surface temperature, one can estimate the gain factors of radiative feedback for all-sky and clear-sky conditions, and also that of cloud radiative forcing (Forster and Gregory, 2006). Comparing the gain factors thus obtained with those obtained from climate models, one can identify any systematic biases of modeled radiative feedback relative to observations. The information resulting from such comparisons should be very useful for validating and improving the models used for the projection of global change.

The global mean fluxes of longwave and solar radiation at the top of the atmosphere have been the subject of analysis by Forster and Gregory (2006). Because the satellite observations of these fluxes were available for only a short time period, Inamdar and Ramanathan (1998) and Tsushima and Manabe (2001, 2013) examined the annual variation of the fluxes rather than their long-term change. Nevertheless, these studies underscore the merit of long-term satellite monitoring of the TOA fluxes for the study of climate sensitivity.

CHAPTER 7

Glacial-Interglacial Contrast

In the early 1970s, a meeting was held at GFDL to explore possible collaboration between climate modelers and paleoclimatologists. This was a time when a consortium of earth scientists had undertaken an ambitious project to reconstruct the state of the Earth's surface at the last glacial maximum (LGM), which occurred approximately 21,000 years ago. At the meeting, John Imbrie of Brown University, one of the leaders of this project, convinced Manabe that the climates of the geologic past would be one of the most promising avenues of research for climate modelers. Their conversation began a long-term research project that continued during much of Manabe's research career. In the early 1980s, Broccoli joined his group after graduate school at Rutgers University and began some of the modeling studies of glacial climate that will be presented in this chapter.

Paleoclimatologists and paleoceanographers have devoted a great deal of effort to reconstructing of the states of the ocean, land surface, and atmosphere at the LGM based upon the analysis of sediments from oceans and lakes, air bubbles trapped in ice cores, and other geologic signatures. Given the glacial-interglacial difference in surface temperature, continental ice sheets, and concentrations of greenhouse gases, attempts have been made to estimate the sensitivity of climate using climate models. In this chapter, we shall describe some of these attempts, searching for the most likely value of climate sensitivity.

The Geologic Signature

In the early 1970s, Imbrie and Kipp (1971) published a study that opened the way for the quantitative analysis of LGM climate. Using multiple regression analysis to relate the taxonomic composition of planktonic biota in deep sea

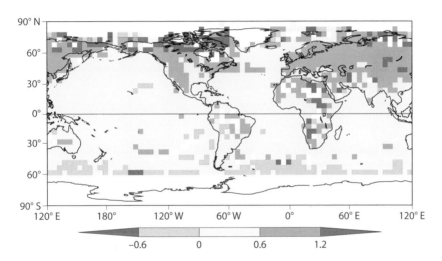

COLOR PLATE 1 Geographic distribution of the observed 25-year (1991–2015) mean surface temperature anomaly (°C) defined as the deviation from the 30-year base period (1961–90). The map was constructed using the historical surface temperature data set HadCRUT4 that was compiled by the Climate Research Unit of the University of East Anglia and Hadley Centre of the UK Meteorological Office (Morice et al., 2012). Note that in the Southern Ocean, poleward of 60° S, the anomaly is not shown because few data are available in winter. From Stouffer and Manabe (2017).

(a) Simulated

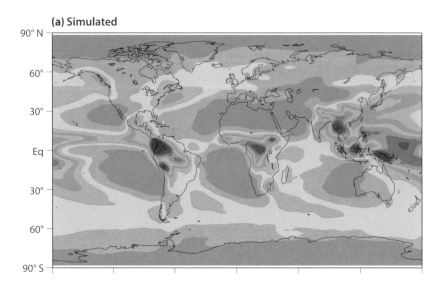

(b) Observed

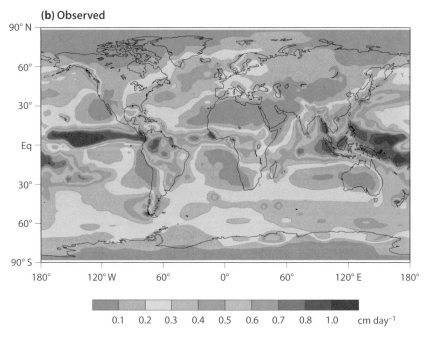

0.1 0.2 0.3 0.4 0.5 0.6 0.7 0.8 1.0   cm day⁻¹

COLOR PLATE 2 Geographic distribution of the annual mean rate of precipitation (cm day⁻¹): (a) simulated; (b) observed (Legates and Willmott, 1990). From Wetherald and Manabe (2002).

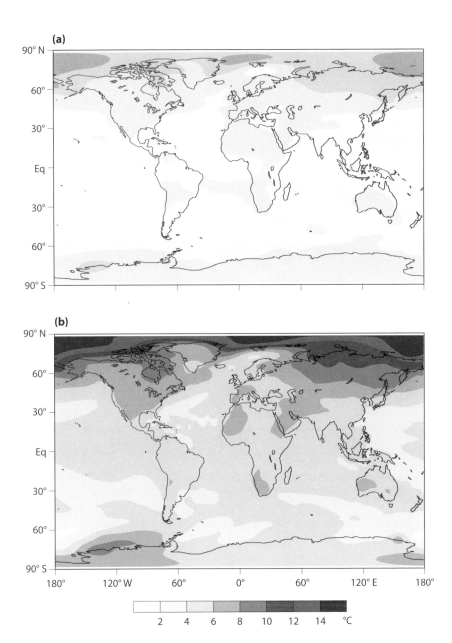

COLOR PLATE 3 Geographic distribution of the simulated change in surface air temperature (*a*) from the preindustrial period to the middle of the twenty-first century, when the CO_2 concentration has doubled, and (*b*) in response to the quadrupling of atmospheric CO_2. From Manabe et al. (2004b).

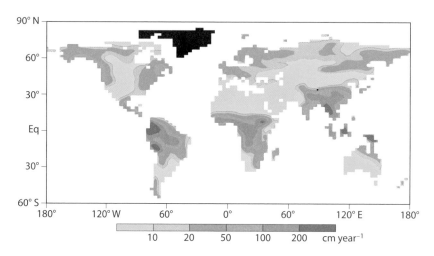

COLOR PLATE 4 Geographic distribution of the annual mean rate of runoff obtained from the control experiment (1×C) in cm year⁻¹. Black shading indicates ice-covered surfaces.

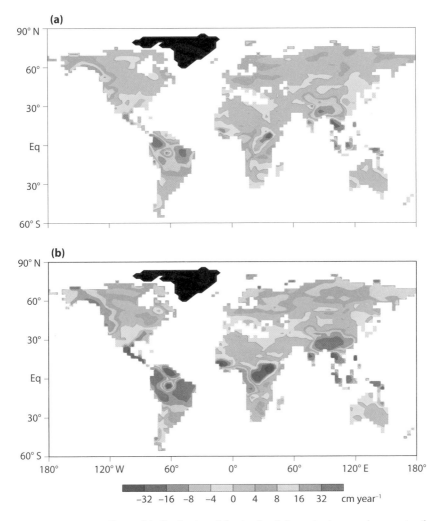

COLOR PLATE 5 Geographic distribution of the simulated change in the annual mean rate of runoff (cm year^{-1}) in response to (*a*) doubling and (*b*) quadrupling of the atmospheric concentration of CO_2. Black shading indicates ice-covered surfaces. From Manabe et al. (2004b).

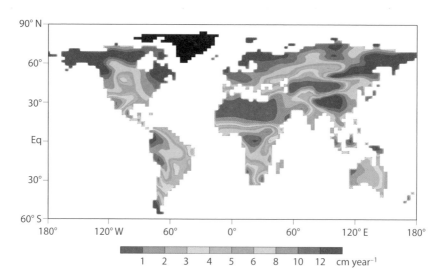

COLOR PLATE 6 Geographic distribution of annual mean soil moisture (cm) obtained from the control run (1×C). Soil moisture is defined as the difference between the total amount and the wilting point of water in the root zone of soil. Black shading indicates ice-covered surfaces. From Wetherald and Manabe (2002).

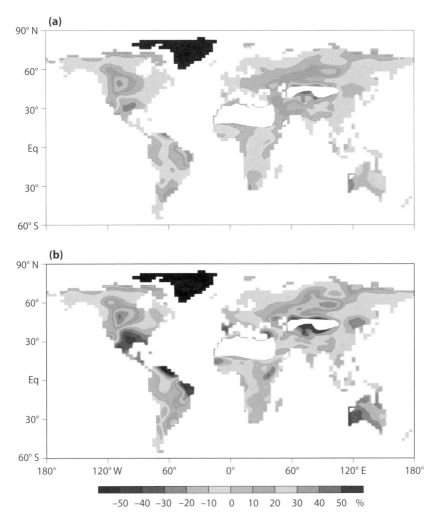

COLOR PLATE 7 Geographic distribution of the percentage change in annual mean soil moisture in response to (*a*) doubling and (*b*) quadrupling of the atmospheric concentration of CO_2. The
percentage change is defined as the percentage of the time-mean soil moisture obtained from the
control experiment. It is not shown in extremely arid regions such as the Sahara and Central Asia,
where soil moisture is less than 1 cm, as shown in plate 6. In these regions, soil moisture is so small
that its percentage change has no physical significance. Black shading indicates ice-covered surfaces.
From Manabe et al. (2004b).

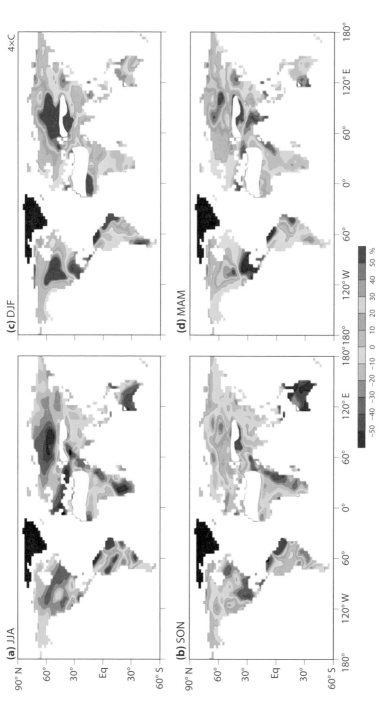

COLOR PLATE 8 Geographic distribution of the percentage change of three-month-mean soil moisture of the model in response to the quadrupling of atmospheric CO_2: (*a*) June–July–August (JJA); (*b*) September–October–November (SON); (*c*) December–January–February (DJF); (*d*) March–April–May (MAM). The percentage change is defined as the percentage of the time-mean soil moisture obtained from the control experiment. It is not shown in extremely arid regions such as the Sahara and Central Asia, where soil moisture is less than 1 cm in plate 6. In these regions, soil moisture is so small that its percentage change has no physical significance. Black shading indicates ice-covered surfaces. From Manabe et al. (2004b).

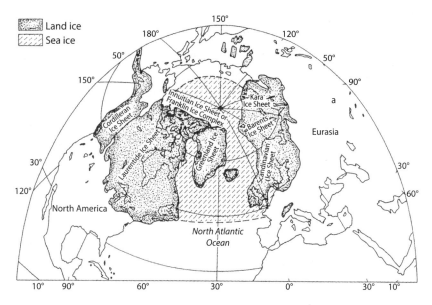

FIGURE 7.1 Distribution of continental ice sheets at the last glacial maximum. Modified from Denton and Hughes (1981).

sediments to sea surface temperature (SST), they found a close relationship between these two variables. Applying the relationship thus obtained, it was possible to translate the abundances of different types of planktonic biota preserved in deep sea sediments into estimates of past SST. Encouraged by the very promising results obtained from this method, Imbrie and colleagues embarked upon the Climate: Long-range Investigation, Mapping, and Prediction Project (CLIMAP; CLIMAP Project members, 1976, 1981) in order to reconstruct the state of the Earth's surface at the LGM.

The most dramatic features of the Earth at the LGM were the great ice sheets that covered large portions of Northern Hemisphere continents (figure 7.1). Although ice cover increased in many areas, the largest accumulation occurred in northeastern North America (Laurentide ice sheet) and northwestern Europe (Fennoscandian ice sheet), with smaller increases in western North America (Cordilleran ice sheet), European Russia, the Alps, the southern Andes, and West Antarctica.

One of the important products of CLIMAP is the reconstruction of these massive LGM ice sheets undertaken by Denton and Hughes (1981). Taking into consideration the outer margins of the ice sheets as indicated by geologic signatures such as moraines, they determined the shape of the ice sheets that would be in dynamic equilibrium. Over ice-free continental surfaces, CLIMAP reconstructed the distribution of vegetation at

LGM from the taxonomic composition of pollen, using multiple regression analysis (Webb and Clark, 1977). The reconstructions of continental ice sheets and vegetation have been useful to climate modelers for determining the distribution of continental surface albedo during the LGM.

An important factor that influences the climate of the LGM is the concentration of atmospheric greenhouse gases such as carbon dioxide, methane, and nitrous oxide. The analysis of air bubbles found in ice cores reveals that the CO_2-equivalent concentration of these greenhouse gases at the LGM was about two-thirds of the preindustrial level (e.g., Chappellaz et al., 1993; Neftel et al., 1982), indicating that the greenhouse effect of the atmosphere was substantially smaller at the LGM than at present. Along with the massive continental ice sheets and extensive snow cover and sea ice that collectively reflect a large fraction of incoming solar radiation, the reduction of the concentration of greenhouse gases is also responsible for making the climate of the LGM much colder than the present. The reduction in greenhouse gases is particularly important in the Southern Hemisphere, where changes in albedo were not as extensive. Here we describe some of the early attempts to simulate the LGM climate using climate models and evaluate its implications for the sensitivity of climate.

Simulated Glacial-Interglacial Contrast

The first attempts to simulate the LGM climate using atmospheric GCMs were undertaken by Williams et al. (1974) and Gates (1976). These studies prescribed surface boundary conditions, including SSTs, based on geologic reconstructions of the LGM. Gates (1976) was the first to use the LGM distributions of SST, sea ice, and albedo of snow-free surface reconstructed by CLIMAP. In his study, the distribution of temperature at the continental surface was obtained as an output of the experiment. He found that the simulated temperature at many continental locations was broadly consistent with the temperature that CLIMAP estimated using various proxy data, such as the taxonomic composition of pollen in lake sediments (Webb and Clark, 1977). The agreement, however, does not necessarily imply that the model had realistic sensitivity, because surface temperature over the continents was closely constrained by prescribing the surface temperature of the surrounding oceans based on the CLIMAP reconstruction.

Hansen et al. (1984) conducted a similar numerical experiment using a version of the GISS atmospheric GCM. The distribution of seasonally varying SST in their control simulation was prescribed based upon modern observations. They also performed an LGM simulation using the LGM distribution of SST as reconstructed by CLIMAP, as Gates did in his study. They computed the difference in the net TOA flux of outgoing radiation between the two experiments and found that the heat loss from the net TOA flux is larger by about 1.6 W m^{-2} in the LGM than the modern simulation. The excess heat loss indicated that the atmosphere-surface system of the model was attempting to cool further than the prescribed SST of LGM would allow. Hansen et al. interpreted this radiative imbalance as an indication either that their model was too sensitive (at ~4.2°C to CO_2 doubling) owing to overly large radiative feedback, or that the CLIMAP estimates of LGM SST used as a lower boundary condition were too warm.

The study of Hansen et al. attempted to infer the sensitivity of climate from the difference in the TOA flux of radiation between the two experiments, in which the LGM and modern distributions of SST were prescribed according to the CLIMAP reconstruction and modern observations, respectively. A more direct way to determine the climate sensitivity of a model would be to simulate the glacial-interglacial SST difference and compare it with the difference as determined by the CLIMAP reconstruction. The first attempt to pursue this approach was made by Manabe and Broccoli (1985). Here we describe the results they obtained and evaluate their implications for climate sensitivity.

The model used for their study was an atmosphere/mixed-layer-ocean model that was developed by Manabe and Stouffer (1980) as described in chapter 5. Two versions of this model were used. The fixed cloud (FC) version is the original version of the model, in which the distribution of cloud was prescribed and cloud feedback was absent. In the variable cloud (VC) version, the distribution of cloud was allowed to change and cloud feedback was in operation. These two versions of the model were originally constructed for the study of cloud feedback that was later published by Wetherald and Manabe (1988), as described in chapter 6. The sensitivity of the VC version to CO_2 doubling was 4°C and is much larger than the ~2°C sensitivity of the FC model. This large difference is attributable not only to the absence of cloud feedback in the FC version, but also to the difference in the strength of the sea ice albedo feedback in the Southern Hemisphere, where surface air temperature is substantially colder in the VC than in the FC version of the model. Irrespective of the specific causes of the difference in sensitivity, these two versions of the model were used in an effort to identify the sensitivity of climate that is more consistent with the glacial-interglacial difference in SST.

Manabe and Broccoli (1985) performed two sets of numerical experiments, using these two versions of the model. Each set of experiments included a present-day control run and a simulation of the LGM climate. All simulations started from an initial condition of an isothermal atmosphere at rest, but different boundary conditions were specified for the LGM and control simulations. In the LGM simulation, the atmospheric greenhouse gas concentration, surface elevation, and ice sheet and land-sea distributions were prescribed according to the ice core and CLIMAP reconstructions. (Changes in the Earth's orbital parameters were not included, even though they have been identified as drivers of glacial-interglacial climate variations, because the parameters at the LGM happen to be similar to those of the present.) Owing in no small part to the absence of the large thermal inertia of the deep ocean below the mixed layer, it took only several decades for the model to closely approach a state of equilibrium. Throughout the course of the time integrations, the seasonal cycle of incoming solar radiation, the albedo of snow-free surface, and the CO_2-equivalent concentration of greenhouse gases were held fixed. The glacial-interglacial contrast of SST was then obtained as the difference between the simulated state of the LGM and that of the present.

Latitudinal profiles of the glacial-interglacial difference in zonal mean SST obtained from the FC and VC versions of the model (figure 7.2) can be compared with the zonal mean temperature difference obtained by CLIMAP. In this comparison, SST in ice-covered regions is defined as the temperature of the water at the bottom surface of the sea ice. Although the SST difference is larger in VC than FC at most latitudes, it is difficult to say which version of the model is closer to CLIMAP. This is because the difference between the CLIMAP reconstruction and either version of the model is much larger than the difference between the two versions of the model at many latitudes. For example, in low latitudes, the glacial-interglacial differences in SST obtained from both versions are substantially larger than the difference obtained by CLIMAP. In the middle latitudes of the both hemispheres, they are comparable in magnitude. In high latitudes of the Northern Hemisphere, poleward of 60° N, they are much smaller than the difference reconstructed by CLIMAP. Nevertheless, we found it encouraging that the latitudinal profiles of the glacial-interglacial SST difference obtained from the two versions of the model resemble broadly the profile obtained by CLIMAP.

Figure 7.3 illustrates schematically how the latitudinal profile of CLIMAP SST changes between the LGM and the present. As shown in figure 7.3a, the SST is relatively high in low latitudes at both the LGM and the present and decreases gradually with increasing latitudes up to the outer margin of sea ice, where it is at the freezing point of seawater (−2°C). The difference between the LGM and the present, shown in figure 7.3b,

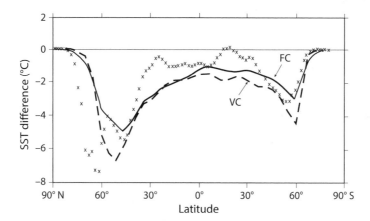

FIGURE 7.2 Latitudinal profiles of zonally averaged annual mean SST difference between the LGM and the present obtained from the FC and VC versions of the atmosphere/mixed-layer-ocean model. The crosses indicate the SST differences (average of January and July) obtained by CLIMAP. From Manabe and Broccoli (1985).

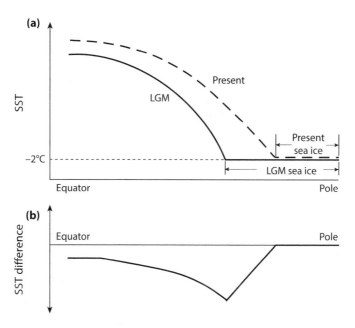

FIGURE 7.3 Schematic diagram illustrating (*a*) the correspondence between sea ice coverage and the latitudinal profile of SST at LGM and at present, and (*b*) the latitudinal profile of the glacial-interglacial SST difference (LGM − present).

increases gradually with latitude, reaching a maximum at the LGM sea ice margin. Poleward of the LGM sea ice margin, the SST difference decreases sharply until it reaches zero over the high latitude ocean, which was covered by sea ice at the LGM and is also ice covered at present. The latitudinal profile of the glacial-interglacial SST difference described in this schematic resembles the profiles in figure 7.2, particularly in the Southern Hemisphere, where the distributions of SST and sea ice are more zonal than in the Northern Hemisphere.

In the numerical experiments performed here, the glacial-interglacial SST difference depends upon three factors: an expansion of continental ice sheets with high surface albedo, a reduction of CO_2-equivalent concentration of greenhouse gases, and an increase in the albedo of the snow-free surface. In order to evaluate the individual contributions of these changes to the glacial-interglacial difference in SST, Broccoli and Manabe (1987) performed additional numerical experiments using the FC version of the model. In each experiment, they changed one factor at a time, thereby evaluating the contribution of each change to the total glacial-interglacial SST difference.

Figure 7.4 illustrates the latitudinal distributions of the individual contributions obtained from this set of experiments. The expansion of the ice sheets has a large impact on SST in the Northern Hemisphere, whereas it is small in the Southern Hemisphere, where glacial-interglacial difference in continental ice extent is small. As noted by Manabe and Broccoli (1985), the damping of SST perturbation through interhemispheric heat exchange is much weaker than the in situ radiative damping in each hemisphere and has little effect upon the magnitude of the perturbation in middle and high latitudes. The effect of the lowered greenhouse gas concentrations is comparable in magnitude between the two hemispheres, though it is substantially larger in the Southern Hemisphere. The contribution from land albedo is relatively small in both hemispheres. Expressed in another way, the expanded ice sheets have the largest impact in the Northern Hemisphere, followed by the lowered concentration of greenhouse gases. In the Southern Hemisphere, on the other hand, the reduced greenhouse gas concentration is mainly responsible for the SST difference. Averaged over the entire globe, both expanded continental ice sheets and reduced greenhouse gases have a substantial impact on global mean SST, whereas changes in land albedo over ice-free areas have only a minor effect on a global basis.

The geographic distribution of the glacial-interglacial SST difference obtained from the VC version of the model is compared with the CLIMAP reconstruction in figure 7.5 as an example. In general, the model simulates reasonably well the broad-scale pattern of the SST difference. For example,

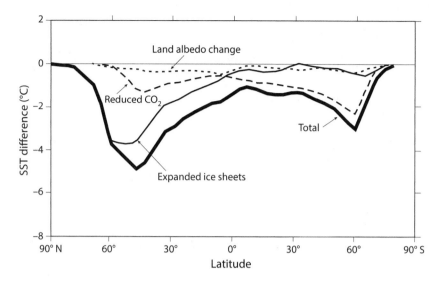

FIGURE 7.4 Latitudinal distributions of the individual contributions from expanded continental ice sheets, reduced atmospheric CO_2, and changes in albedo of ice-free land surface to the total glacial-interglacial SST difference (°C) obtained from the FC version of the atmosphere/mixed-layer-ocean model. From Broccoli and Manabe (1987).

regions of relatively large SST difference appear in the zonal belt of the Southern Ocean and in the northern North Atlantic Ocean, where the LGM sea ice margins are located, consistent with the schematic diagram shown in figure 7.3.

Upon close inspection, however, one can identify many differences between the model simulation and the CLIMAP reconstruction. In the Southern Hemisphere, for example, the glacial-interglacial difference in SST is large in the zonal belt around 60° S in the model simulation, while it is large around 50° S in the CLIMAP reconstruction. In the northern North Atlantic, the SST difference is large in both the model simulation and the CLIMAP reconstruction. However, the area of large difference extends farther northward along the Scandinavian coast in the CLIMAP reconstruction. As shown schematically in figure 7.3, the glacial-interglacial SST difference reaches a maximum along the equatorward margin of sea ice at the LGM. Thus, it is likely that both of the discrepancies are at least partially attributable to the failure of the atmosphere/mixed-layer-ocean model used here to place realistically the location of the LGM sea ice margin.

In the CLIMAP reconstruction shown in figure 7.5b, the glacial-interglacial difference in SST has small positive values over extensive regions of the Pacific Ocean on both sides of the equator, suggesting that the sea surface was warmer at the LGM than at present in these regions.

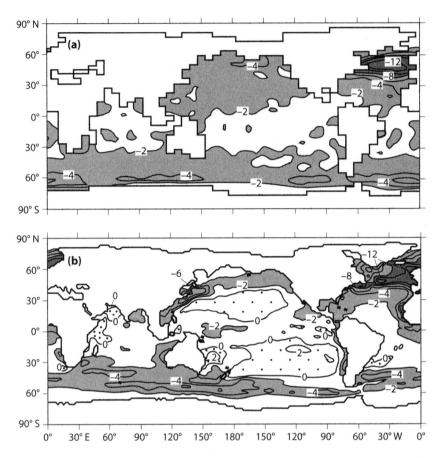

FIGURE 7.5 Geographic distribution of annual mean SST difference (°C) between the LGM and present (LGM − present): (*a*) SST difference obtained from the VC version of the model; (*b*) CLIMAP estimates (average of February and August). From Manabe and Broccoli (1985).

This reconstructed warming is quite counterintuitive and warrants close scrutiny. Examining the source of the data used by CLIMAP, Broccoli and Marciniak (1996) found that very few sediment core data are available in the regions of positive SST difference, and SST in these regions was often determined through subjective hand-contouring interpolation from the surrounding regions. Realizing that the SST difference obtained by CLIMAP may not be as reliable in these regions, they recomputed the zonal mean SST difference using the data only at those locations where sediment core data were available. In the tropical latitudes (30° N–30° S), the zonally averaged glacial-interglacial SST difference thus obtained is −1.8°C, which is much larger than the zonal mean SST difference of −0.6°C originally reconstructed by CLIMAP. It is, however, similar to the zonal mean temperature differences of −1.6°C and −2.0°C obtained from the FC and VC versions of the model, respectively.

Since the publication of the CLIMAP study, many studies have suggested that tropical SST at the LGM was much lower than the CLIMAP reconstruction. For example, based upon the isotopic analysis of corals off Barbados, Guilderson et al. (1994) suggested that the tropical SST was colder by ~5°C than it is today. Beck et al. (1992) estimated tropical SST based upon the high positive correlation between SST and the strontium/calcium ratio in living corals. They found that the LGM tropical SST was colder than it is today by ~5°C, in agreement with the result obtained by Guilderson et al. These estimates of the tropical SST are much lower than the 1.8°C revised estimate of Broccoli and Marciniak.

On the other hand, Crowley (2000) found it very difficult to believe that tropical SST was so cold at LGM, wondering whether it would be possible for most corals to live in the tropical ocean at such low temperatures. He speculated that, if tropical SSTs were 5°C colder than the present, most corals would have been on the edge of or below the present level of habitability, and the taxonomic composition of plankton would have been quite different from what is indicated in the analysis conducted by CLIMAP. Therefore, he concluded that the SST difference between the LGM and present must be substantially less than the large values obtained by Guilderson et al. (1994) and Beck et al. (1992).

An alternative approach for the reconstruction of SST involves the use of alkenone molecules, which are produced by plankton and are preserved in marine sediments. For example, Brassell et al. (1986) found a strong correlation between temperature and the ratio of two types (diunsaturated and triunsaturated) of alkenone molecules. During the past few decades, many attempts have been made to estimate the glacial SST using the relationship between this ratio and SST. It appears that, in the coastal regions of the Atlantic and Indian oceans, where alkenone estimates were made, the modest reductions in SST reconstructed by this technique are not substantially different from those obtained by CLIMAP.

Improving the VC version of the model described above, Broccoli (2000) made a renewed attempt to simulate the distribution of SST at the LGM. The computational resolution of his model is twice as high as that of the FC or VC version of the model used for the study described above. In addition, he used the so-called Q-flux method, developed by Hansen and his collaborators at GISS, described in chapter 6. The application of this method allowed the geographic distributions of SST and sea ice in the control experiment to be more realistic than those obtained from either of the earlier FC or VC versions of the model. The sensitivity of this new version

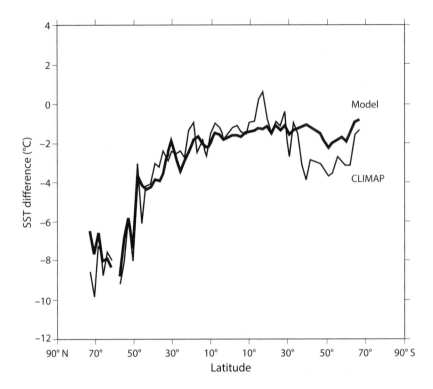

FIGURE 7.6 Latitudinal profiles of zonally averaged annual mean SST difference (°C) between LGM and present (LGM – present), from CLIMAP and the model of Broccoli (2000). Zonal averages were computed as the arithmetic average of SSTs at those locations where sediment-core data were available for CLIMAP. From Broccoli (2000).

of the model was 3.2°C, which is the median of the models evaluated for the IPCC *Fifth Assessment Report* (Flato et al., 2013).

Figure 7.6 compares the simulated and reconstructed profiles of the zonally averaged glacial-interglacial SST difference. Both profiles were determined using the SSTs from only those locations where CLIMAP sediment cores were available. As this figure shows, the agreement between the simulated and reconstructed profile is excellent, not only in the tropics but also in middle and high latitudes of the Northern Hemisphere. The agreement is also excellent up to ~35° S in the Southern Hemisphere. Poleward of 40° S, however, the SST difference simulated by the model is substantially smaller than the difference obtained from the CLIMAP sediment-core data. The discrepancy is the subject of further discussion in chapter 9, which evaluates the equilibrium response of the coupled atmosphere-ocean model to a reduction in the atmospheric concentration of CO_2. It suggests that the discrepancy is attributable to an absence of the interaction between the upper layer and the deep layer in the Southern Ocean of the atmosphere/mixed-layer-ocean model that Broccoli used in his study.

Given that the sensitivity of the model is 3.2°C, the similarity to the CLIMAP reconstruction would support the possibility that the sensitivity of the actual climate is not substantially different from 3°C. This result appears to be consistent with the results of Hansen et al. (1984) (discussed earlier in this chapter), which suggests that the actual sensitivity is likely to be less than the 4.2°C sensitivity of his model. Putting together the results obtained here with those from the preceding chapter, one can speculate that the sensitivity of the actual climate is likely to be about 3°C.

Developments in paleoceanography since the Broccoli study was completed at the turn of the millennium have led to additional approaches for estimating LGM SSTs. For example, the magnesium/calcium (Mg/Ca) ratio in shells of marine microorganisms has been used to estimate LGM tropical temperatures. Lea (2004) summarized results from a number of studies using Mg/Ca as well as alkenones to reconstruct SSTs, and concluded that tropical oceans cooled by 2.8°C ± 0.7°C at the LGM. Such cooling is somewhat larger than the CLIMAP estimates, but not as large as the early estimates from corals described previously. The Multiproxy Approach for the Reconstruction of the Glacial Ocean Surface project estimated an LGM reduction of tropical mean SST of 1.7°C (MARGO Project members, 2009), which is similar to the CLIMAP reconstruction as interpreted by Broccoli and Marciniak (1996). Annan and Hargreaves (2013) estimated a tropical SST cooling of 1.6°C ± 0.7°C.

These more recent reconstructions of tropical SST remain broadly consistent with a global climate sensitivity to CO_2 doubling of 3°C. A comprehensive analysis of reconstructions of past temperatures and radiative forcing from throughout Earth's geologic history (PALAEOSENS Project members, 2012) estimated a range of 2.2°C–4.8°C for the actual climate sensitivity. Although the uncertainty associated with this estimate remains larger than would be desired, the sensitivity of the model employed by Broccoli (2000) lies close to the center of this range.

CHAPTER 8

The Role of the Ocean in Climate Change

The Thermal Inertia of the Ocean

In the preceding chapters, we have discussed the so-called equilibrium response—that is, the total response of climate to thermal forcing, given a sufficiently long time. In this chapter, we shall discuss the time-dependent response of climate to thermal forcing, using GCMs of the coupled atmosphere-ocean-land system. To begin, we shall identify the factors that control the time-dependent response of climate to thermal forcing, using a simple zero-dimensional energy balance model of the coupled atmosphere-ocean-land system introduced by Schneider and Thompson (1981).

As discussed in chapter 6, the heat balance of the atmosphere-ocean-land system is maintained between the net incoming solar radiation and outgoing longwave radiation at the top of the atmosphere. Let us consider the surface-atmosphere system that is in thermal equilibrium. If the system is heated, it is expected that the temperature of the Earth's surface and that of the overlying atmosphere would increase with time. The prognostic equation of the global mean surface temperature may be expressed as follows:

$$C \, \partial T'/\partial t = Q - \lambda \, T', \qquad (8.1)$$

where T' is the deviation of the global mean surface temperature from its initial value in thermal equilibrium, C is the effective heat capacity of the system, Q is the thermal forcing applied to the system, and t denotes time. λ is the so-called feedback parameter defined by equation (6.5) in chapter 6. It denotes the strength of the radiative damping that operates on the perturbation in the global mean surface temperature through the net outgoing radiation from the top of the atmosphere.

Suppose that a thermal forcing, Q, is abruptly applied to a system (i.e., "switched on") that is initially in thermal equilibrium. In other words, $Q = 0$ for $t < 0$, and $Q = Q_0$ for $t \geq 0$, as shown in figure 8.1b. In this case, the universal solution of equation (8.1) may be expressed in nondimensional form as follows:

$$(T'/T'_\infty) = 0 \text{ for } t < 0$$

$$(T'/T'_\infty) = [1 - \exp(-t/\tau)] \text{ for } t \geq 0, \tag{8.2}$$

where T'_∞ is the value of T' when $t = \infty$, Q_0 is the thermal forcing applied, and τ is the time constant of the response, which is expressed by:

$$\tau = C/\lambda. \tag{8.3}$$

The time constant is often called the e-folding time, and it denotes the time $t = \tau$, or $t/\tau = 1$, when $(1 - 1/e)$, or about 63%, of the total equilibrium response has been achieved, with $1/e$, or about 37%, remaining. This response is shown in figure 8.1a, which illustrates the normalized surface temperature change from its initial value. In other words, the time constant indicates the time needed to achieve approximately two-thirds of the total response and is often used as an indicator of the length of delay in response. As equation (8.3) implies, the time constant, τ, is proportional to the effective heat capacity of the system, C. For the climate system, almost all of the effective heat capacity resides in the ocean. The time constant is also inversely proportional to the strength of the radiative damping that operates on the global-scale perturbation of surface temperature. We will begin our evaluation of the role of the ocean in climate change by using this simple model to see how the ocean delays the response of surface temperature to thermal forcing.

In order to use equation (8.3) to estimate the time constant of the model response to switch-on thermal forcing, it is necessary to estimate not only the effective heat capacity of the system, C, but also the magnitude of the feedback parameter, λ. As discussed in chapters 6 and 7, the most likely value of the sensitivity of climate, defined as the equilibrium response of the global mean surface temperature to CO_2 doubling, is ~3°C. Assuming that the sensitivity of climate is 3°C and the thermal forcing of the CO_2 doubling is ~4 W m^{-2}, one can estimate an approximate value of the feedback parameter using equation (6.7) in chapter 6. The feedback parameter thus obtained is ~1.3 W m^{-2}K^{-1}. In order to estimate the global mean heat capacity of the earth system, C, it is assumed that the land surface has

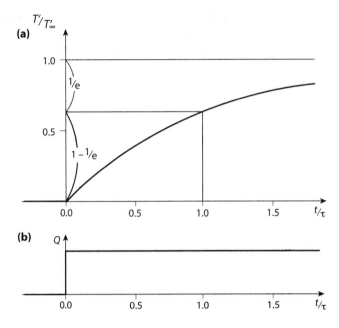

FIGURE 8.1 Temporal variation of (*a*) the normalized global mean surface temperature anomaly, and (*b*) switch-on thermal forcing. *Q*, switch-on thermal forcing; *t*, time; *T'*, global mean surface temperature anomaly from its initial condition; T'_∞, its total equilibrium response; τ, the time constant of the response.

essentially no heat capacity. Assuming also that the mixed-layer ocean has a thickness of about 70 m and does not exchange heat with the ocean below, one can estimate that the global mean heat capacity of the system is ~2 × 10⁸ J m⁻². Using these values of λ and *C* and equation (8.3), the time constant, τ, is about five years, indicating a very short time scale of ocean response.

The simple model of the ocean mixed layer used in the above calculation neglects the heat exchange between the mixed layer and the deeper ocean. Levitus et al. (2000) found that the heat content of the ocean changed substantially between the mid-1950s and the mid-1990s, not only in the well-mixed surface layer, but also at depths from 300 to 1000 m. They also found that heat penetrated below 1000 m in the Atlantic Ocean. Their results clearly indicate that temperature has increased in the deep subsurface layer as well as the surface layer of ocean. It is therefore likely that the actual time scale of the full ocean is substantially longer than five years.

The importance of heat exchange between the surface layer and the deeper ocean was anticipated by Thompson and Schneider (1979), Hoffert et al. (1980), and Hansen et al. (1981). Using simple models of the coupled atmosphere-ocean-land system that explicitly treat the heat exchange between the surface layer and the deep ocean, they made initial attempts to simulate the time-dependent response of the coupled atmosphere-ocean-land system to the gradually increasing thermal forcing that results from an upward trend in atmospheric CO_2 concentration. They found that the delay in the warming is small during the first few decades of their experiments. However, the length of delay increases as heat is mixed downward from the surface layer to the deeper layer of the ocean, thereby increasing the effective thermal inertia of the system.

We will look more closely at the work of Hansen et al. (1981) as an example of these early studies to explore the effect of the deep ocean on the transient response of climate. They constructed a set of 1-D globally averaged vertical column models, in which a radiative-convective model of the atmosphere was coupled with an ocean model. In the first version of the model, the ocean is treated as a well-mixed surface layer and an underlying deep layer in which heat diffuses vertically. The coefficient of vertical diffusion used here is 10^{-4} $m^{-2}s^{-1}$ and is similar to the value that Munk (1966) inferred from the vertical distributions of temperature and salinity in the ocean. The second version has only the surface layer of the ocean without the deep layer; in the third version, the heat capacity of the ocean is zero.

Using the three versions of the model identified above, Hansen and his collaborators computed the time-dependent response of the global mean surface temperature to a gradual increase in the atmospheric concentration of CO_2 during the 120-year period from 1880 to 2000. Figure 8.2 illustrates, for the three versions of the model, how the global mean surface temperature changes from its initial value. In the version of the model in which ocean has no heat capacity, it takes about 100 years for the global mean surface temperature to increase by 0.5°C. In the version in which the ocean is treated as only a well-mixed surface layer, it takes ~5 additional years (i.e., the time constant obtained from the simple zero-dimensional model expressed by equation [8.1]) for the same 0.5°C warming to be realized. In the most complete version of the model, in which the ocean has both a surface layer and an underlying deep layer, it takes ~10 additional years for the same 0.5°C warming to be realized. In short, the heat exchange with the deep ocean delays the warming by about 15 years, or about three times as long as the delay caused by the well-mixed surface layer alone. The results presented here clearly indicate how the delay in surface temperature response increases as heat penetrates into the deeper layer of the ocean.

In addition to the numerical experiments described above, Hansen et al. performed an additional set of experiments, using the version of the model that has not only the surface layer but also the underlying deep ocean.

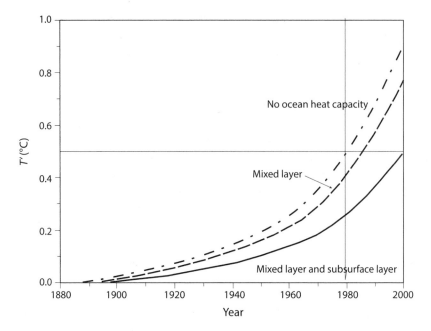

FIGURE 8.2 Temporal variations of the global mean surface temperature anomalies.

In these experiments, the model is forced by variations in CO_2, volcanic aerosols that reflect incoming solar radiation, and solar irradiance over the 100-year period between 1880 and 1980. Among the three types of radiative forcing applied, the estimate of the CO_2 forcing is the most reliable, followed by that of volcanic aerosols. The estimate of solar variability is highly conjectural and the least reliable, as noted by the authors. In view of the large differences in reliability among the three types of radiative forcing, the model was forced by the three sets of radiative forcing: CO_2 only, CO_2 plus volcano, and CO_2 plus volcano plus Sun. Figure 8.3 compares the time series of the global mean surface temperature obtained from these three model simulations with observed temperatures. It is quite encouraging that the agreement between the simulated and observed time series improves with each additional thermal forcing. However, one should not take too seriously the slight improvement that results from the addition of solar variability, which is highly uncertain.

As a natural extension of the study described above, Hansen et al. (1988) conducted a landmark study of global warming using a 3-D model of the coupled atmosphere-ocean-land system. The model covered the entire globe and had a realistic distribution of oceans and continents. It was

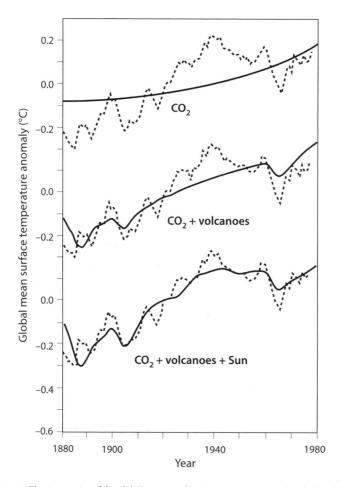

FIGURE 8.3 The time series of the global mean surface temperature anomalies obtained from the 1-D vertical column model thermally forced by CO_2, CO_2 + volcanoes, and CO_2 + volcanoes + Sun. From Hansen et al. (1981).

constructed by combining a GCM of the atmosphere, a relatively simple land surface model, and an ocean model that consists of a surface layer and a deep, vertically diffusive layer that underlies the surface layer. The coefficient of vertical diffusion in this deep layer varied geographically and was determined empirically from the relationship between the penetration of transient inert tracers and local water column stability.

Using this model, they computed the climate change in response to the best available estimates of radiative forcing due to the changes not only in carbon dioxide, but also in other trace constituents of the atmosphere, such as methane, nitrous oxide, CFCs, and sulfate aerosols of volcanic origin.

Encouraged by the success in simulating the global mean change of surface temperature during the last half of the twentieth century, they extended their computation into the twenty-first century. They found that the magnitude of the warming was far from uniform geographically and tended to be larger over the continents than over the ocean. The warming increased with increasing latitudes in both hemispheres, in qualitative agreement with the equilibrium response of their atmosphere/mixed-layer model obtained earlier (Hansen et al., 1984). Through this experiment, they demonstrated convincingly that their 3-D model of the coupled atmosphere-ocean-land system was a very powerful tool for predicting global warming. Hansen presented the highlights of the results of this study at a hearing of the US Congress held in 1988. His testimony received wide attention and had a far-reaching impact on public awareness of global warming.

In the oceanic component of the global GISS model described above, the vertical transport of heat in the deep subsurface layer of ocean is parameterized as vertical eddy diffusion. In the actual ocean, however, heat is transported vertically not only by turbulence and convection, but also by the large-scale circulation. These processes interact closely with one another. In order to predict global-scale changes of the Earth system, it is therefore desirable to construct a model that combines a general circulation of the atmosphere with a model of the ocean in which turbulence, convection, and large-scale circulation are incorporated explicitly. The initial attempt to construct such a model was made by Manabe and Bryan (1969) at GFDL, combining an atmospheric model with the GCM of the ocean constructed by Bryan and Cox (1967). Although their model had a limited computational domain with an idealized distribution of land and sea, it simulated successfully the salient features of the observed distribution of temperature and precipitation in the coupled system. Encouraged by their success, they began to use such models to simulate and study global warming (Bryan et al., 1982, 1988).

In their 1988 study, for example, Bryan et al. employed a coupled model with a computational domain that consisted of three identical sectors extending over both Northern and Southern hemispheres. Each sector was bounded by two meridians that were 120° apart. The fraction of ocean in each latitude belt was specified to be similar to the actual fraction. In the zonal belt between 55° and 60° S, for example, the model ocean was zonally connected, mimicking the circumpolar Southern Ocean. Using the model thus constructed, they obtained the response of the coupled system to switch-on thermal forcing, in which the atmospheric CO_2 concentration doubled abruptly at the beginning of the experiment and remained unchanged thereafter. They found that the SST increased with increasing latitude in the Northern Hemisphere, whereas it hardly changed in the in the high latitudes of the Southern Hemisphere. As will be shown later in this chapter, a similar interhemispheric asymmetry was also evident in the distribution of surface temperature change obtained from a global coupled model with realistic geography. This asymmetry is in contrast to the result obtained by Hansen et al. (1988), in which polar amplification of warming occured in both hemispheres. The difference between the two results suggests that it may be desirable to incorporate explicitly the effect of heat transport by ocean currents into a coupled model.

The initial attempts to construct a fully coupled global GCM with realistic geography at GFDL and the National Center for Atmospheric Research (NCAR) took place in the 1970s (Manabe et al., 1979; Washington et al., 1980). By the late 1980s, groups at both institutions had published results from more realistic global warming experiments, in which the atmospheric CO_2 concentration was increased gradually (Stouffer et al., 1989; Washington and Meehl, 1989). In the remainder of this chapter, we will discuss the role of the ocean in controlling the distribution of surface temperature change, based upon an analysis of the experiments conducted by Manabe et al. (1991, 1992). We will begin by describing briefly the structure of the global atmosphere-ocean GCM developed at GFDL.

The Coupled Atmosphere-Ocean Model

Figure 8.4 illustrates the structure of the global coupled atmosphere-ocean GCM that Stouffer et al. (1989) used in their study of global warming. For simplicity of identification, we will hereafter refer to this GCM as the "coupled model." The model consisted of an atmospheric GCM, an oceanic GCM, and simple models of the heat and water budgets at the continental surface. The model had realistic geography, in contrast to earlier coupled model versions that employed a sector domain with highly idealized geography (e.g., Manabe and Bryan, 1969).

The basic structure of the atmospheric GCM was described in chapter 4. It computes the rates of change of wind, temperature, and specific humidity, using the equation of motion, thermodynamic equation, and continuity equation of water vapor, respectively. Over the continents, the atmospheric GCM is coupled with a simple model of heat and water budget, as described in chapter 4. The oceanic GCM (Bryan and Lewis, 1979) computes the rates of change of ocean currents, temperature, salinity, and sea ice thickness using the equation of motion, the thermodynamic equation, the prognostic equation of salinity, and a simple model of sea ice, respectively. The sea

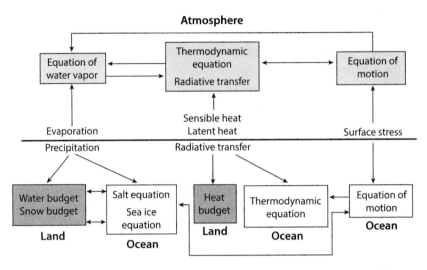

FIGURE 8.4 Diagram depicting the structure of the coupled atmosphere-ocean model. From Stouffer et al. (1989).

ice model is similar to the so-called thermodynamic model described in chapter 5, although the sea ice is this version moves with the surface ocean currents. These atmospheric and oceanic GCMs interact with each other, exchanging heat, water vapor, and momentum at the interface.

Because of the limited capability of the computers available in the late 1980s, when this experiment was performed, the computational resolution of the model is much lower than the coupled models that are currently available. For example, the atmospheric GCM had nine vertical finite difference levels. The model employed the so-called spectral method, in which the horizontal distributions of the predicted variables were represented by both spherical harmonics (15 associated Legendre functions for each of the Fourier components) and grid point values. The oceanic GCM had a regular grid system with $4.5° \times 3.75°$ (latitude $\times$ longitude) spacing and 12 unevenly spaced vertical finite difference levels. The thickness of the top finite difference layer was 50 m, representing a vertically isothermal ocean mixed layer.

When the time integration of a coupled atmosphere-ocean model is started from a realistic initial condition, its climate often drifts away from a realistic state because the model is not a perfect representation of the real climate system. For obvious reasons, such drift can distort the time-dependent response of climate to thermal forcing. To reduce this drift, a method called "flux adjustment" was employed in the study presented here. In the next section, we will briefly describe this method and explain why it is effective for predicting and evaluating climate change. Further details of the method can be found in the paper by Manabe et al. (1991).

Initialization and Flux Adjustment

In order to prevent the climate drift described above, the initial condition for the time integration of the coupled model was obtained from the separate time integrations of the atmospheric and oceanic components of the coupled model. The initial condition of the atmospheric component was an isothermal and dry atmosphere at rest, with a realistic distribution of seasonally dependent SST prescribed at the oceanic surface. The time integration of this atmospheric model was performed over a period of 12 years. After a few years, the model atmosphere attained a quasi-equilibrium, in which the seasonal variation of the atmospheric state repeated itself quite well from one year to the next. The state of the model atmosphere thus obtained was used as the atmospheric initial condition for the time integration of the coupled model.

The initial condition for the oceanic component of the model consisted of a vertically well-mixed surface layer with realistic horizontal distributions of temperature, salinity, and sea ice thickness, and a deep layer with constant temperature and salinity. From this starting point, the preliminary time integration of the ocean component was continued over a period of approximately 2400 years. Throughout the course of the integration, the above variables were relaxed toward seasonally and geographically varying observed values in the surface layer, with a relaxation time of 50 days. Temperature, salinity, and velocity were predicted below the surface layer. Toward the end of the time integration, there was no systematic trend in the oceanic state except in the very deep layer of ocean, where temperature was changing very slowly. This quasi-equilibrium oceanic state was used as the initial condition for the time integration of the coupled model. The latitude-depth profile of the zonal mean ocean temperature thus obtained is compared with the observed profile in figure 8.5. Although the deep water temperature is too warm by 1°C–2°C, the latitude-height profile of the zonal mean temperature simulated by the model resembles the observed profile.

The atmospheric and oceanic states that were obtained from these separate preliminary integrations of the atmospheric and oceanic components of the model were combined to produce the initial condition for the time integration of the coupled model. Owing to imperfections of these component models, the seasonally varying horizontal distributions of the heat

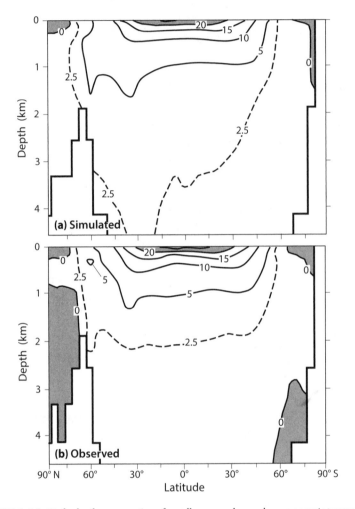

FIGURE 8.5 Latitude-depth cross-section of zonally averaged annual mean oceanic temperature: (a) simulated temperature at the end the 2500-year integration of the oceanic component of the coupled atmosphere-ocean model; (b) observed temperature from Levitus (1982).

and water fluxes at the oceanic surface obtained from the atmospheric component differ from those obtained from the oceanic component. These differences make it likely that the oceanic state of the coupled model would drift if the interfacial flux obtained from the atmospheric component were to be imposed upon the oceanic component. In order to prevent the drift, the interfacial fluxes of heat and water obtained from the atmospheric component of the model were modified by an amount equal to the difference between the fluxes obtained from the oceanic and atmospheric components in their separate integrations. After being modified in this way, the fluxes were then imposed on the oceanic component as the coupled model was integrated. Although the flux adjustment values depended upon season and geographic location, they did not change from one year to the next and were thus independent of the simulated state of the surface ocean. By applying the flux adjustments as described above, temperature, salinity, and sea ice thickness at the oceanic surface of the coupled model fluctuated around realistic values and hardly drifted systematically with time. When a perturbation experiment was conducted by applying a radiative forcing to the model, identical spatially and seasonally varying adjustments were applied to the perturbed simulation as well as the control so that it did not distort artificially the difference in the atmosphere-ocean fluxes between the control and perturbed simulations.

Maintaining a realistic distribution of temperature and sea ice thickness is important for estimating the sensitivity of climate. As discussed in chapters 4 and 6, the sensitivity of climate to a given thermal forcing depends critically upon the temperature at the Earth's surface, mainly because the strength of the albedo feedback of snow and sea ice intensifies with decreasing surface temperature. Owing to the application of the flux adjustment, the distributions of temperature and sea ice thickness were realistic, making the model well suited for estimating climate sensitivity.

The use of flux adjustment has been effective for reducing markedly the artificial drift that can occur in coupled models. Reducing climate drift is the main reason why this method was employed in the global warming experiments of Manabe et al. (1991, 1992). In our opinion, flux adjustment is likely to be a valuable tool for predicting climate change as well as for estimating climate sensitivity. There are other examples of experiments in which flux adjustment has been, or is likely to be, effective for predicting climate change. We will briefly discuss a few such experiments.

As pointed out in chapter 4, tropical cyclone activity depends critically upon the distribution of SST in low latitudes. Because a coupled ocean-atmosphere model with flux adjustment usually maintains a realistic distribution of SST in the absence of thermal forcing, it is well suited for the near-term prediction of tropical cyclone activity. Thus, it is not surprising that Vecchi et al. (2014) found that the flux adjustment approach was effective for improving forecasts of seasonal hurricane frequency in the Atlantic Ocean. In a similar vein, Manganello and Huang (2009) were able to improve substantially retrospective predictions of the Southern Oscillation, using a relatively simple heat flux adjustment scheme that is constant with time.

As shown by Manabe and Stouffer (1997), for example, the intensity of the Atlantic Meridional Overturning Circulation (AMOC) depends critically upon the horizontal distributions of salinity and temperature in the upper layer of the ocean. Because a coupled model using flux adjustment maintains realistic distributions of these variables, it has been very useful for simulating the AMOC (Manabe and Stouffer, 1988) and its multidecadal oscillations (Delworth et al., 1993). For this reason, flux adjustment is likely to be effective for improving decadal predictions of this deep overturning circulation and its response to climate change on multidecadal time scales.

Increasing efforts have been made to improve parameterization of various sub-grid-scale processes, such as cloud microphysics and sea ice dynamics. The parameterizations of these processes have become very detailed, introducing many additional parameters. Thus, it has been very difficult to tune these parameters such that the geographic distributions of key variables (e.g., temperature, salinity, and sea ice thickness) are realistic at the oceanic surface. We hope that the method of flux adjustment described above can be used to complement model tuning as the complexity of parameterizations increases in the future.

Global Warming Experiments

The global warming experiment of Manabe et al. (1991, 1992) utilized two time integrations of the coupled model that were started from the initial condition described above. In the control run, in which the atmospheric CO_2 concentration was held fixed at a value of 300 ppmv, the global mean SST fluctuated around a realistic value with little systematic trend, indicating that the flux adjustment was effective for preventing the systematic drift of temperature. In the global warming run, on the other hand, the CO_2 concentration was increased at a compounded rate of 1% per year, which happened to be similar to the rate of increase of the CO_2-equivalent concentration of well-mixed greenhouse gases around the time when this experiment was performed. The solid line in figure 8.6 indicates how the global mean surface temperature of the coupled model increased in this run at a gradually increasing pace owing to the gradual increase in atmospheric CO_2 concentration. By the seventieth year, when the CO_2 concentration had doubled, the temperature had increased by about 2.5°C.

As discussed at the beginning of this chapter, warming at the oceanic surface is reduced and delayed owing to the thermal inertia of the deep ocean that underlies the surface mixed layer. In order to evaluate the magnitude of the reduction in response and the length of delay, the time-dependent response of the global mean surface temperature of the coupled

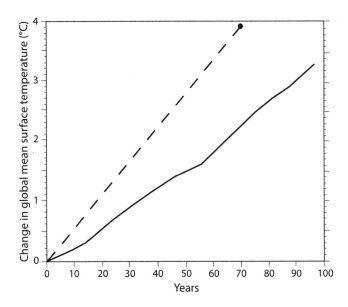

FIGURE 8.6 The time-dependent response of the global mean surface air temperature (°C) of the coupled atmosphere-ocean model to the gradual increase in atmospheric CO_2 concentration at the (compounded) rate of 1% $year^{-1}$ (solid line). The black dot represents the equilibrium response of the global mean surface air temperature of the atmosphere/mixed-layer-ocean model to doubling of the atmospheric CO_2 concentration. The dashed line connecting the origin of the graph and this dot approximates the almost linear growth of the equilibrium response of the global mean surface temperature with time. From Manabe et al. (1991).

model is compared in figure 8.6 with that of an atmosphere/mixed-layer-ocean model. Although the atmospheric components of the two models are identical to each other, the oceanic component of the atmosphere/mixed-layer model does not have the deep ocean layer. Instead, the Q-flux method described in chapter 7 was used, in which the heat exchange between the mixed layer and the deep ocean is prescribed such that the geographic distributions of SST and sea ice thickness are realistic. An identical heat flux is prescribed in the CO_2-doubling run, implicitly assuming that heat exchange between the mixed layer and the deep ocean remains unchanged. Although it would be better to estimate the equilibrium response by using the coupled atmosphere-ocean model described earlier, the atmosphere/mixed-layer model is used to circumvent the computational cost of running the coupled model to quasi-equilibrium. Owing to the Q-flux method, the distributions of both SST and sea ice thickness are realistic in the control run, as they are in the coupled model with flux adjustment.

The equilibrium response of global mean surface air temperature to CO_2 doubling obtained from the atmosphere/mixed-layer model is about

3.9°C, which is substantially larger than the 2.5°C response of the coupled model at the time of CO_2 doubling. The equilibrium response is plotted in figure 8.6 as a black dot at the seventieth year, when the CO_2 concentration has doubled. A straight dashed line is drawn between the origin of the graph and the black dot, implicitly assuming that the equilibrium response increases linearly as CO_2 concentration increases exponentially at the compounded rate of 1% per year. This assumption is justified because the greenhouse effect of the atmosphere is proportional to the logarithm of the CO_2 amount in the atmosphere, as discussed in chapter 1. The horizontal distance between the dashed and solid lines represents the lag of the time-dependent response relative to the equilibrium response. As this figure shows, the lag is zero at the beginning but increases gradually with time. By the seventieth year, when the CO_2 concentration has doubled, the lag is about 30 years. This result implies that the effective thermal inertia of the ocean increases gradually as heat penetrates into the deep subsurface layer of ocean, thereby delaying the warming at the Earth's surface.

So far, we have examined the time-dependent response of the global mean surface air temperature. Now we shall examine the geographic distribution of the surface temperature response at the time that CO_2 concentration has doubled. Figure 8.7a shows that surface air temperature increases almost everywhere with the exception of a very small area north of the Weddell Sea, where it decreases slightly. In agreement with the result obtained by Hansen et al. (1988), for example, the warming is larger over the continents than over the oceans, particularly in the Northern Hemisphere, where the continents occupy a larger fraction of surface area than they do in the Southern Hemisphere.

One of the most notable features of this figure is the large interhemispheric asymmetry in the magnitude of the warming at the Earth's surface. In the Northern Hemisphere, the warming tends to increase with increasing latitudes and is at a maximum over the Arctic Ocean. In contrast, it is relatively small in high southern latitudes. As discussed in chapter 5, the large warming in high northern latitudes is attributable mainly to the poleward retreat of the sea ice and snow cover that reflect a large fraction of incoming solar radiation. On the other hand, the small warming in high southern latitudes is attributable mainly to deep convective mixing of heat not only near the Antarctic coast but also in a very extensive region in the Southern Ocean, as will be discussed later in this chapter. The geographic pattern of the surface air temperature change presented above is broadly similar to the multimodel pattern of the surface temperature change projected by a majority of the models used for the IPCC *Fifth Assessment Report* (see lower left panel of figure 12.41 in Collins et al. [2013]).

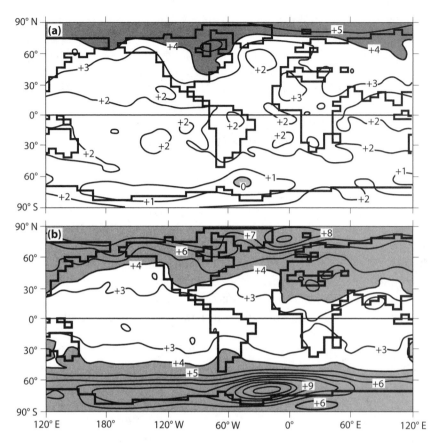

FIGURE 8.7 Geographic distribution of (*a*) the change in surface air temperature of the coupled atmosphere-ocean model realized by around the seventieth year of the global warming experiment, when the atmospheric CO_2 concentration has doubled; (*b*) the equilibrium response of surface air temperature of the atmosphere/mixed-layer-ocean model in response to the doubling of the atmospheric CO_2 concentration. The change in (*a*) is the difference between the 20-year (years 60–80) mean surface air temperature of the global warming run and the 100-year mean temperature obtained from the control run, in which the atmospheric CO_2 concentration is held fixed at the standard value (300 ppmv). Note that surface air temperature represents the temperature at the lowest finite-difference level at the height of about 70 m. Units are °C. From Manabe et al. (1991).

The geographic distribution of the time-dependent response of surface air temperature obtained from the coupled model described above may be compared with that of the equilibrium response of the atmosphere/mixed-layer-ocean model shown in figure 8.7b. As expected, the time-dependent response is smaller than the equilibrium response, indicating that the former lags behind the latter almost everywhere on the globe. It is quite notable, however, that the equilibrium response of surface air temperature increases with increasing latitudes in both hemispheres. This is in sharp contrast to the large interhemispheric asymmetry evident in the time-dependent response.

Inspecting figure 8.7, one finds that the land-sea difference in the magnitude of the warming in middle and high latitudes of the Northern Hemisphere is evident not only in the time-dependent response but also in the equilibrium response. This suggests that it is attributable not only to the thermal inertia of the ocean that delays the warming but also to other factors. As noted by Manabe et al. (1992), the land-sea difference in the equilibrium response appears during much of the seasonal cycle. In winter, when snow cover often extends to midlatitudes, the albedo feedback is much stronger over the continents than over the ocean and is mainly responsible for the land-sea difference in the magnitude of warming. In other seasons, a similar land-sea contrast in the magnitude of the warming exists because the evaporative removal of heat is more effective from the perpetually wet ocean surface than from the relatively dry continents. In the time-dependent response of the coupled model, the thermal inertia of ocean is also responsible for delaying the warming, particularly in certain oceanic regions, thereby enhancing the land-sea difference in warming as described below.

Figure 8.8 illustrates the geographic distribution of the ratio of the time-dependent response to the equilibrium response described above. The ratio is less than 0.4 in a wide zonal belt over the Southern Ocean and dips below 0.2 near the coast of Antarctica. This implies that the delay of response is more than 40 years in the Southern Ocean and is longer than 60 years near the Antarctic coast. In the northern North Atlantic Ocean between Greenland and the west coast of Europe, the ratio is also less than 0.4, indicating that the delay there is longer than 40 years. Over the remainder of the world, including both continental and oceanic regions, the ratio is about 0.7–0.8, corresponding to a delay of 15–20 years. In short, the delay over the Southern Ocean and the northern North Atlantic Ocean is longer than the globally averaged delay of ~30 years, whereas the delay over the rest of the world is generally shorter than the global average.

The time-dependent response of surface air temperature in the coupled model (figure 8.7a) can be compared with the trend of the observed surface temperature change during the past several decades (see plate 1) as determined from the anomaly of the 25-year mean surface temperature for the period 1991–2015, relative to a 30-year base period (1961–90). Because of the removal of short-term fluctuations through time averaging, these anomalies can be regarded as an indicator of the long-term trend of surface temperature during the past half-century, when global-scale warming has been most pronounced (see figure 1.1).

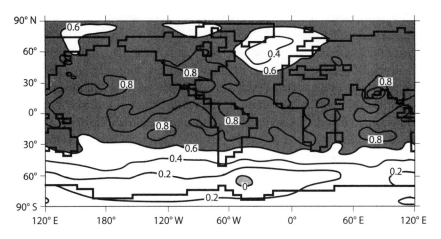

FIGURE 8.8 Geographic distribution of the ratio of the time-dependent response of surface air temperature of the coupled atmosphere-ocean model (shown in figure 8.7a) to the equilibrium response of surface air temperature of the atmosphere/mixed-layer-ocean model (shown in figure 8.7b). From Manabe et al. (1991).

Although the observed temperature anomalies are spotty, owing partly to sampling limitations, they are positive over much of the globe. This implies that surface temperature has increased almost everywhere during the past several decades. In the Northern Hemisphere, the magnitudes of the anomalies are relatively large over both the Eurasian and North American continents and increase with increasing latitude. In the Southern Ocean, however, the anomalies are small poleward of 50° S, with both positive and negative sign. This is in sharp contrast to high northern latitudes, where the anomalies usually have large positive values. It is quite encouraging that the geographic pattern of the observed surface temperature anomalies described above resembles that of the transient response of surface air temperature shown in figure 8.7a. The similarity between the two patterns underscores the possibility that the coupled model contains the basic physical processes that control the large-scale distribution of global warming at the Earth's surface. The commentary by Stouffer and Manabe (2017) provides further discussion of this subject.

One should note here that the thermal forcing used in this study involved increases in the CO_2-equivalent concentration of greenhouse gases. Changes in other forcing agents such as anthropogenic aerosols, solar irradiance, and volcanic aerosols are ignored. The similarity between the simulated and observed patterns suggests that the geographic distribution of surface temperature change may not depend critically upon the pattern of thermal forcing.

Arctic Ocean (JJA)   **Southern Ocean** (DJF)

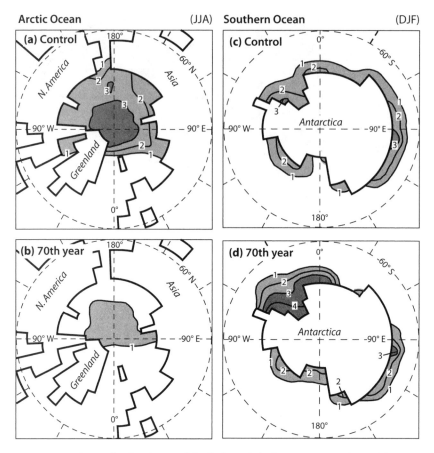

FIGURE 8.9 Geographic distribution of the thickness (m) of summer sea ice obtained from the global warming experiment. (*a*) Initial and (*b*) 70th-year thickness of Arctic sea ice averaged over the three-summer-month period of June, July, and August (JJA). (*c*) Initial and (*d*) 70th-year thickness of Antarctic sea ice averaged over the three-summer-month period of December, January, and February (DJF). The initial thickness represents the average over the 100-year period of the control run, in which the CO_2 concentration is held fixed. The 70th-year thickness represents the average over the 20-year period (years 60–80) of the global warming run, in which the CO_2 concentration increases gradually at the compounded rate of 1% year^{-1}. From Manabe et al. (1992).

The interhemispheric asymmetry of the response manifests itself not only in SST but also in the geographic distribution of sea ice thickness in summer at the time of CO_2 doubling (figure 8.9). Both areal coverage and thickness of sea ice in the Arctic Ocean and surrounding regions decrease markedly. Although it is not shown here, a substantial reduction in thickness also occurs in winter. In the Southern Ocean, on the other hand, the thickness and areal coverage of summer sea ice increases in the Weddell Sea and its immediate vicinity but does not change systematically in other regions. Qualitatively similar changes also occur in winter (not shown), when sea ice expands toward low latitudes. Annually averaged, the thickness and areal coverage of sea ice decrease markedly in the Arctic and subpolar oceans, whereas these variables hardly change in the Southern Ocean, with the exception of the Weddell and Ross seas, where sea ice thickness increases as global warming proceeds. The interhemispheric asymmetry of sea ice change simulated by the model appears to be broadly consistent with that of surface temperature described earlier in this chapter.

Changes in annual mean sea ice extent have differed considerably between the two hemispheres during the past several decades, during which comprehensive observations are available from satellite microwave sensors. Vaughan et al. (2013) presented the time series of the annual mean sea ice extent over the Arctic and Antarctic regions in the IPCC *Fifth Assessment Report*. These time series are reproduced in figure 8.10. Over the Arctic, the annual mean extent of sea ice has decreased at a rate of 3.8% per decade. In the Antarctic, on the other hand, sea ice has been increasing at a rate of +1.5% per decade. Although decreases in ice have occurred in the Bellingshausen and Amundsen seas, located to the west of the Antarctic Peninsula, increases have occurred in other sectors (Vaughan et al., 2013). The difference in the observed long-term trend of sea ice extent between the two hemispheres is in qualitative agreement with the results obtained from the coupled model.

So far, we have shown that the increase in surface air temperature is delayed markedly in certain oceanic regions, such as the northern North Atlantic and the Southern Oceans. In the remainder of this chapter, we will attempt to determine why the warming is delayed greatly in these regions, based upon the analysis conducted by Manabe et al. (1991, 1992).

The Atlantic Ocean

In the upper layers of the Atlantic Ocean, saline and warm water moves northward into the vicinity of Iceland, where it is cooled by frigid air that is advected eastward from Canada and Greenland in winter. As the water cools, it becomes dense, inducing deep convection. Consequently, water sinks near Greenland and moves southward at depth along the east coast of North and South America. This global-scale overturning circulation was called the "Great Ocean Conveyor" by Broecker (1991) and is illustrated schematically in figure 8.11 following Gordon (1986). The total flow of water involved is about 20 million $m^3 s^{-1}$, or 20 Sverdrups (Sv) in the units used by oceanographers. This flow is about 20 times as large as the total discharge

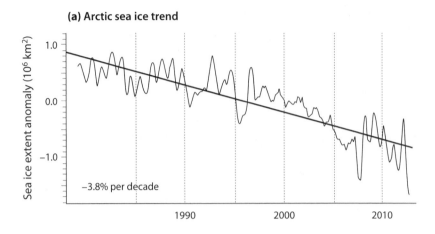

(a) Arctic sea ice trend

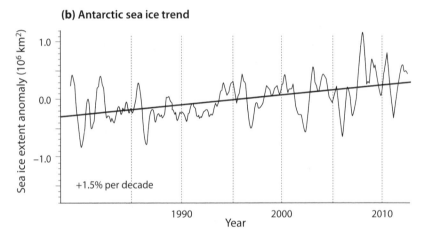

(b) Antarctic sea ice trend

FIGURE 8.10 Time series of the observed sea ice extent anomaly relative to the mean of the entire period (1980–2005) for (*a*) the Northern Hemisphere and (*b*) the Southern Hemisphere, based upon passive microwave satellite data. The linear trend for each hemisphere is indicated by the thick line. From Vaughan et al. (2013).

from all rivers of the world. Transporting warm upper ocean water to the northern North Atlantic and Nordic seas (e.g., the Norwegian and Greenland seas), the conveyor is partially responsible for maintaining a relatively warm climate over the northern North Atlantic and western Europe.

The broad-scale features of the Atlantic meridional overturning circulation described above were simulated reasonably well by the coupled model, as noted by Manabe and Stouffer (1988, 1999). Because of the vertical mixing of heat due to deep convection, the effective thermal inertia of the ocean is very large in the narrow sinking region near Greenland, delaying

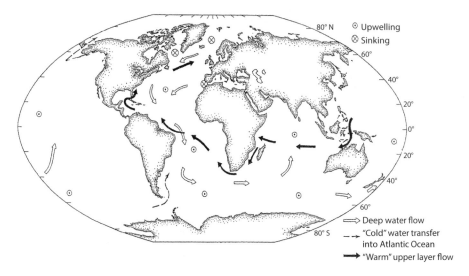

FIGURE 8.11 Global distribution of the overturning circulation. Solid arrows indicate the direction of the upper layer flow; outlined arrows indicate the direction of the flow of deep water. Narrow sinking regions are identified by crosses enclosed by circles, and broad upwelling regions are indicated by dots enclosed by circles. From Gordon et al. (1986).

greatly the warming of the oceanic surface in the region. This is the main reason why the ratio of the time-dependent response to the equilibrium response (figure 8.8) is less than 0.4 off the southeast coast of Greenland, where the sinking region of the ocean conveyor is located.

In addition to the deep vertical mixing, there are other factors that contribute to the delay in the warming in the North Atlantic and its immediate vicinity. The structure and strength of the meridional overturning circulation can be depicted by plotting its zonally averaged streamfunction for the Atlantic basin (figure 8.12). The rate of the overturning at the beginning of the global warming experiment was ~17 Sv (figure 8.12a), a bit smaller than its observed value. By the seventieth year, when the CO_2 concentration had doubled, it had been reduced to 12 Sv, as indicated by the absence of the 15 Sv contour in figure 8.12b. Due to the slowdown of the overturning circulation, the advection of warm and saline near-surface water toward the sinking region decreases. Thus the warming is reduced not only over the narrow sinking region of the conveyor, where the ratio of the transient to the equilibrium response is less than 0.4, but also in the surrounding regions, which include southeastern Greenland, Labrador, and northern and western Europe, where the ratio is less than 0.6 (figure 8.8).

In order to find the reason for this weakening of the conveyor, Manabe et al. (1991) conducted a detailed analysis of their global warming

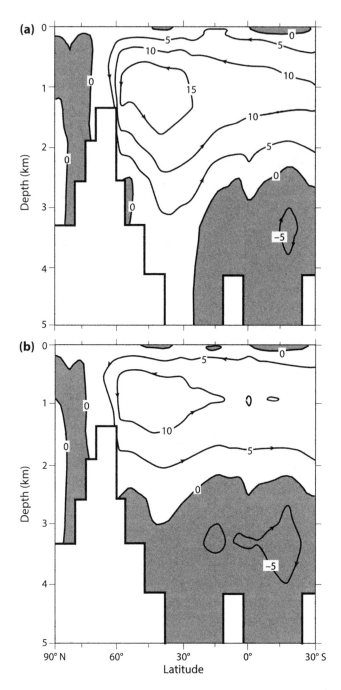

FIGURE 8.12 Streamfunction of the time-mean meridional overturning circulation zonally averaged over the Atlantic Ocean at (*a*) the beginning, and (*b*) the ~70th year of the global warming experiment that was performed using the coupled atmosphere-ocean model. Units are Sverdrups (10^6 m^3 s^{-1}). From Manabe et al. (1991).

experiments. They found that the slowdown of the conveyor was attributable mainly to a reduction of surface salinity in the northern North Atlantic Ocean. When temperature increases in the troposphere owing to global warming, the absolute humidity of the air tends to increase, as previously noted, enhancing the poleward transport of water vapor by extratropical cyclones from the subtropics toward high latitudes, as will be described in chapter 10. The increase in poleward transport, in turn, results in an increase in precipitation in high latitudes and is responsible for the reduction of surface salinity not only in the Arctic Ocean but also in the surrounding oceans, such as the northern North Atlantic. This process accounts for the relatively fresh surface water with low density that caps the sinking region of the conveyor, thereby weakening the overturning circulation in the Atlantic Ocean.

The density of surface water decreases not only because of the reduction of surface salinity but also because of the increase in surface temperature. Although both of these changes weaken the conveyor, the former appears to have a much larger impact than the latter in the experiments presented here. Gregory et al. (2005) found that, in most of the climate models they examined, the change in surface temperature weakens the conveyor by at least about 20%. The remaining weakening is attributable to the change in surface salinity. They show that the change in freshwater supply and its effect on surface salinity varies greatly among models, and the variations are partly responsible for the large intermodel differences in the magnitude of the weakening.

So far, observational evidence for a systematic weakening of the Atlantic overturning circulation has been inconclusive. As noted by Delworth et al. (1993), the intensity of the overturning circulation in the coupled model fluctuates on a multidecadal time scale. This is an important reason why the intensity of the overturning circulation did not show a clear decreasing trend until the year 2000 in the global warming experiment of Haywood et al. (1997) that was conducted using the coupled model presented here. A recent study by Caesar et al. (2018) inferred a 15% reduction in overturning strength since the mid-twentieth century by using a "fingerprint" method applied to observed SSTs. In order to detect a long-term trend in the intensity of the overturning circulation more directly, it is desirable to monitor various manifestations of overturning circulation over decadal to centennial time scales.

The global warming experiment described here was extended to multicentennial and millennial time scales (Manabe and Stouffer, 1994). They found that the intensity of the AMOC varied over a multicentury time scale as global warming proceeded. Studies conducted by Manabe and Stouffer

(1993) and Stouffer and Manabe (2003) provide additional analysis of this topic. The role of the overturning circulation in climate change is the subject of the review conducted by Manabe and Stouffer (1999).

The Southern Ocean

The Southern Ocean comprises the southernmost waters of the world ocean and encircles the Antarctic continent. Around 60° S, where this ocean is zonally connected along a latitude circle through Drake Passage, the intense eastward-flowing currents and deep meridional overturning circulation are maintained beneath strong westerly winds. Gill and Bryan (1971) showed how a deep overturning cell is maintained in the ocean, providing unusually strong coupling between the surface layer and the underlying deep ocean.

To a first approximation, ocean currents flow along isobars, maintaining the so-called geostrophic balance between the Coriolis and pressure gradient forces. The most important region where this constraint is broken is the surface mixed layer of ocean, within which the stress exerted by wind is redistributed by turbulence, resulting in a three-way balance among the Coriolis force, pressure gradient force, and surface wind stress. As shown

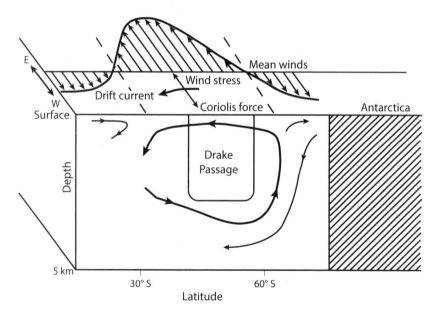

FIGURE 8.13 Schematic illustration of the prevailing surface wind in the Southern Hemisphere, and the zonally averaged overturning circulation at the latitude of the Drake Passage. From Held (1993).

schematically in figure 8.13, which was constructed by Held (1993), the zonal mean wind stress in the near-surface layer of ocean must be balanced by the Coriolis force acting on a nongeostrophic northward drift current. On the other hand, a southward geostrophic return flow develops in the deep ocean, where the Coriolis force acting on the return flow is compensated by the pressure difference across the bottom relief.

If the circumpolar currents were interrupted by a meridional barrier, the southward return flow would develop in the near-surface layer, where the Coriolis force acting on the return flow would be balanced by the zonal pressure gradient between the two sides of the wall. Thus, a shallow overturning cell would be maintained, as in the subtropical oceans that are interrupted by continents. In order to maintain a deep overturning cell, it is therefore necessary that the ocean is zonally connected, as the Southern Ocean is through Drake Passage.

Figure 8.14 illustrates the zonally averaged overturning circulation obtained from the control run of the coupled model. In this figure, one can identify a deep overturning cell, with upwelling around 60° S in the southern

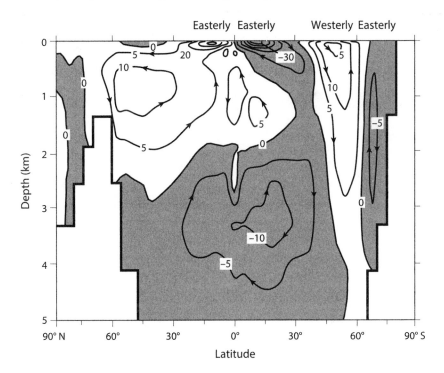

FIGURE 8.14 Streamfunction of the annual mean overturning circulation zonally averaged across all ocean basins. The direction of prevailing surface wind is indicated at the top of the figure. Units are Sverdrups (10^6 m^3 s^{-1}). From Manabe et al. (1991).

flank of the Antarctic Circumpolar Current, and sinking around 40° S in its northern flank. Over the northern flank, the surface westerly winds decrease toward the north and induce a northward drift current that converges at the oceanic surface, pumping water downward. Over the southern flank, on the other hand, the westerlies increase toward the north and induce a drift current that diverges at the oceanic surface, yielding the upwelling of water. Thus, the surface westerlies force a deep overturning circulation often called the "Deacon cell" (Deacon, 1937), which appears as a counterclockwise circulation in a latitude-depth perspective. Poleward of the Deacon cell, a clockwise overturning cell exists beneath the prevailing easterly winds, with deep sinking near the Antarctic coast. The deep overturning circulation has a synergic relationship with deep convection as described below.

In the upwelling region of the Southern Ocean, thin sea ice forms and grows rapidly in winter as it diverges beneath frigid air, yielding, through brine rejection, patches of cold and saline water at the oceanic surface. Because of their higher density, these patches of water sink deeply even though the ocean is slightly stratified, inducing deep convection. Meanwhile, wind-driven upwelling of water prevents the continuous accumulation of the saline, dense water in the deep layer of the ocean, helping to sustain deep convection. In short, the upwelling plays a critically important role in sustaining deep convection in the region.

In the Southern Ocean, deep convection predominates not only in the upwelling region around 60° S as described above, but also in the coastal regions of the Weddell and Ross seas around 75° S, where brine rejection associated with sea ice formation also produces dense water at the ocean surface. This dense water also sinks deeply and moves northward. In addition to the deep convective sinking described above, the exchange of momentum at the surface and bottom of the ocean also contributes to the overturning. Because of deep convection, water mixes deeply in much of the Southern Ocean. This is the main reason why the warming penetrates deeply not only in the upwelling region around 60° S but also in the sinking region around 75° S as described below.

Figure 8.15 illustrates the change in zonal mean temperature of the coupled model realized by the seventieth year of the global warming experiment, when the atmospheric concentration of CO_2 had doubled. As this figure shows, positive temperature anomalies penetrate deeply in the Antarctic Ocean around 60° S and 75° S as well as in the northern North Atlantic Ocean. In these oceanic regions, where deep convection predominates, the effective thermal inertia of ocean is very large, owing in no small part to the deep vertical mixing of heat. Thus, the warming at the oceanic surface is reduced greatly in these regions, as shown in figure 8.8.

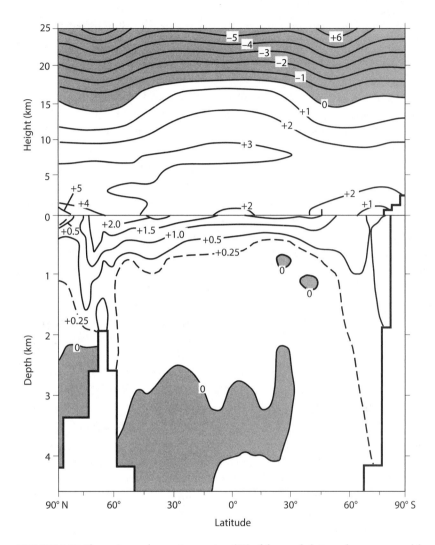

FIGURE 8.15 Change in zonal mean temperature (°C) of the coupled atmosphere-ocean model averaged over the 21-year period centered on the 70th year of the global warming run, when the CO_2 concentration has doubled. From Manabe et al. (1991).

One of the energetic modes of circulation in the Southern Ocean is mesoscale eddies. These eddies have length scales of 10 to 100 km and are too small to be treated explicitly in the coupled model presented here. Nevertheless, they are very active in the circumpolar currents that flow though Drake Passage. Using a GCM of the ocean developed by Gent et al. (1995) that incorporates the parameterization of mesoscale eddies, Danabasoglu et al. (1994) explored the role of these eddies in maintaining the thermal and dynamical structure of the ocean. They found that mesoscale eddies redistribute momentum in the vertical, inducing a mean meridional overturning circulation that flows in the opposite direction from the wind-driven Deacon cell. These two cells canceled each other in their experiment, yielding no residual circulation. Their study suggests that deep overturning circulation may be absent in the actual Southern Ocean, casting serious doubt on the existence of the deep overturning circulation in the circumpolar ocean around 60° S.

Since the publication of the study of Danabasoglu et al., several further studies have been published suggesting that the cancellation between the two cells may not be as complete as it was in their result. Analyzing the results from an eddy-permitting high-resolution ocean model with a grid spacing of 20 km, Henning and Vallis (2005) found that the eddy-induced cell was substantially weaker than the wind-driven Deacon cell in their model. Karsten and Marshall (2002) also found that although the eddy-induced overturning circulation and the Deacon cell oppose each other in observations, the residual mean flow can remain at one-third to one-half of the magnitude of the wind-driven circulation. Based upon their analysis, they suggested that the zero-residual-mean condition may not be wholly appropriate in the real ocean. Their result appears to be consistent with the modeling study conducted by Morrison and Hogg (2013). Using ocean models with various grid sizes (¼°, ⅛°, ¹⁄₁₂°, and ¹⁄₁₆°), Morrison and Hogg evaluated how the strength of the eddy-induced circulation depends upon the resolution of the model and wind stress. They found that decreasing the grid spacing beyond ¹⁄₁₂° had little effect on the strength of the eddy-induced circulation, suggesting that ¹⁄₁₂° resolution may be good enough to resolve mesoscale eddies in the Southern Ocean. Inspecting the result from the ¹⁄₁₂° model for the typical wind stress observed in the Southern Ocean, one finds that the intensity of the eddy-induced circulation is about 40% of that of the Deacon cell, with 60% remaining as the residual circulation. These studies appear to suggest that the deep overturning circulation is maintained in the Southern Ocean despite the compensation between the wind-driven cell and the eddy-induced circulation.

In order to evaluate the performance of the coupled model in simulating deep convective mixing, Dixon et al. (1996) attempted to simulate the downward penetration of CFC-11 (trichlorofluoromethane, or CCl_3F) in the Southern Ocean during the late twentieth century, using the coupled model. As an example, figure 8.16 compares the simulated and observed change in CFC-11 along the route of an Atlantic cruise made near ~0° longitude. It shows that the coupled model simulates reasonably well the

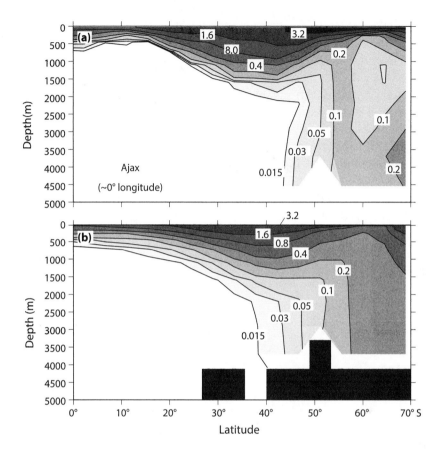

FIGURE 8.16 CFC-11 concentrations (pmol kg^{-1}) along the Ajax prime meridian cruise track: (*a*) measured in October 1983 and January 1984 (Weiss et al., 1990); (*b*) as simulated by the coupled atmosphere-ocean model. From Dixon et al. (1996).

observed deep penetration of CFC-11 poleward of 50° S during the twentieth century. It is encouraging that the model successfully simulates the deep penetration of CFC-11 not only at 0° but also at other longitudes, indicating that deep convective mixing predominates in the southern flank of the Antarctic Circumpolar Current of the coupled model.

Parameterizations of mesoscale eddies, such as that developed by Gent et al. (1995), were incorporated into a majority of the coupled atmosphere-ocean-land models used in the recent IPCC *Fifth Assessment Report* of climate change. However, the specific details of parameterization vary greatly among the models. Nevertheless, it is encouraging that the projected multimodel-mean change in surface temperature (shown in the lower left panel of figure 12.40 of Collins et al. [2013]) is relatively small in the Southern Ocean, in agreement with the results from the global warming experiment presented here.

In summary, the rate of the change in surface air temperature is very small over the Southern Ocean in the global warming experiment presented here. This is attributable mainly to the large thermal inertia of the ocean that delays greatly the response of surface temperature to the gradually increasing concentration of atmospheric CO_2. As discussed in this chapter, deep convection predominates not only in the immediate vicinity of the Antarctic coast, but also in the Southern Ocean around 60° S where it is zonally continuous around the latitude circle, increasing greatly its thermal inertia. For this reason, the polar amplification of warming that predominates in the Northern Hemisphere is practically absent in the Southern Ocean, in agreement with observations.

CHAPTER 9

Cold Climate and Deep Water Formation

In the preceding chapter, we investigated the transient response of climate to a gradual increase in the atmospheric concentration of CO_2. In this chapter, based upon the study conducted by Stouffer and Manabe (2003), we discuss the total equilibrium response of climate to large changes in the atmospheric CO_2 concentration, given a sufficiently long time for the climate to adjust. Although we discussed the equilibrium response to CO_2 doubling in the preceding chapter, that discussion was based on results from an atmosphere/mixed-layer-ocean model, in which heat exchange between the surface layer and the deep ocean is held fixed and does not change with time. Here, we explore the role of the deep ocean in the equilibrium response of climate using the coupled model, in which heat exchange between the surface layer and the deep ocean is incorporated explicitly. Four very long time integrations of the coupled model were performed as described below.

Starting from the realistic initial condition that was discussed in the previous chapter, the time integrations of the coupled model were performed over at least several thousand years, which is long enough for the temperature of deep water to stabilize. The control integration was run with the atmospheric concentration of CO_2 held fixed at the standard value of 300 ppmv. In the 2×C and 4×C integrations, the CO_2 concentration initially increased at a compounded rate of 1% year^{-1} before being held fixed at twice and four times the standard value, respectively. The CO_2 concentration in the ½×C integration initially changed at a compounded rate of −1% year^{-1}, but then was held unchanged at one-half the standard value. The time-varying forcing for each of these four integrations is depicted in figure 9.1. The duration of the time integrations was more than 15,000 years for the control, 4000 years for the 2×C, and 5000 years each for the 4×C and ½×C.

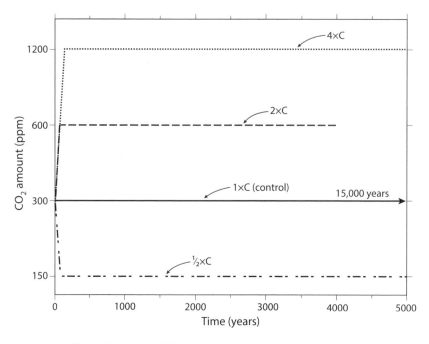

FIGURE 9.1 Temporal variation of the atmospheric concentration of CO_2 (ppmv; logarithmic scale) as prescribed in the coupled atmosphere-ocean model simulations.

Toward the end of all four runs, the global mean ocean temperature at a depth of 3 km was barely changing, indicating that the deep ocean of the model was near a state of thermal equilibrium (figure 9.2). The temperature of deep water toward the end of the four runs was 6.5°C in 4×C, 4.5°C in 2×C, 1°C in 1×C, and −2°C (i.e., the freezing point of seawater at the oceanic surface) in ½×C. A noteworthy aspect of the ½×C run is that the temperature of deep water stabilized earlier than in the other runs, as dense, cold and saline water occupies the deep ocean. The analysis presented here is based on the mean states of the coupled model averaged over the last 100 years of each integration.

As discussed in chapter 1, the atmospheric greenhouse effect increases approximately in proportion to the logarithm of the CO_2 concentration of air. This implies that a doubling of CO_2 concentration from 150 to 300 ppmv exerts approximately the same thermal forcing as a doubling from 300 to 600 ppmv or from 600 to 1200 ppmv, even though the magnitudes of change in CO_2 concentration are quite different from one another. Table 9.1 indicates, however, that the difference in surface temperature between ½×C and 1×C is 7.8°C, and is much larger than the 4.4°C difference between 1×C and 2×C, which in turn is larger than the 3.5°C difference

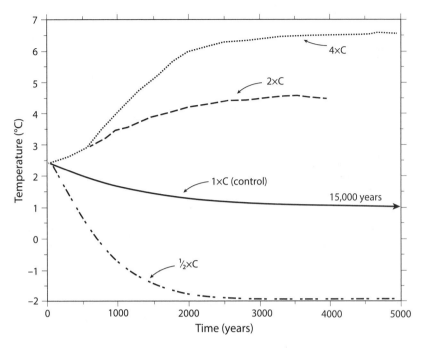

FIGURE 9.2 Temporal variation of global mean deep water temperature (°C) at a depth of 3 km.

TABLE 9.1 *Global mean surface air temperatures at equilibrium in the coupled model*

	Model time integration			
	½×C	1×C	2×C	4×C
Global mean surface temperature (K)	276.4	284.2	288.6	292.1

Temperatures averaged over the last 100-year period of the four time integrations.

between 2×C and 4×C. In short, the equilibrium response of surface temperature to CO_2 doubling decreases with increasing surface temperature, mainly because the strength of the albedo feedback of snow and sea ice decreases as the climate warms, as discussed in chapter 5.

The latitudinal profiles of the zonal mean surface air temperature obtained from the four time integrations are illustrated in figure 9.3a. For close inspection, the differences between the control and other integrations are shown at magnified scale in figure 9.3b. In general, the differences in temperature increase from low to high latitudes, where the albedo feedback of snow and sea ice predominates. Of particular interest is the large

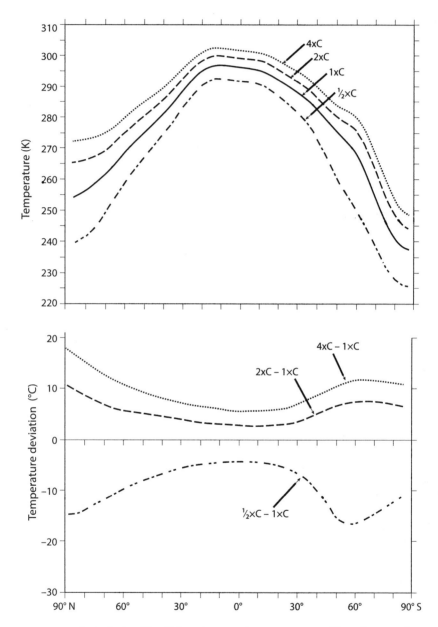

FIGURE 9.3 Latitudinal profiles of (*a*) zonal mean surface air temperature; (*b*) zonal mean surface air temperature deviation from the control (1×C).

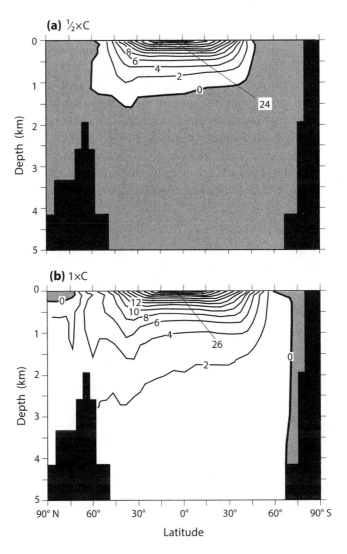

FIGURE 9.4 Latitude-depth distribution of zonal mean temperature (°C): (*a*) ½×C; (*b*) 1×C.

difference in surface air temperature between the control and ½×C in the Southern Ocean around 60° S. This exceptionally large cooling is attributable in no small part to the very extensive perennial sea ice in the Southern Ocean in ½×C. Next, we will describe how the ocean's structure in the ½×C simulation is quite different from its structure in the simulations with higher CO_2 concentrations.

Figure 9.4a illustrates the latitude-depth profile of zonal mean temperature obtained from the ½×C simulation. Dense cold water fills the very thick deep layer of ocean, outcropping at the oceanic surface in the high latitudes of both hemispheres. The temperature of much of this layer is almost isothermal and is close to −2°C, the freezing point of seawater at the oceanic surface, and is substantially lower than the temperature of deep water below the depth of 3 km in the 1×C simulation, which is about +1.5°C. Comparing figure 9.4a with figure 9.4b, which illustrates the profile from the 1×C simulation, one notes that the layer of cold deep water is much thicker in the ½×C simulation. The temperature profile of the ½×C ocean is quite different from that of the 1×C ocean shown in figure 9.4b. For example, the thermocline is shallower and the cold bottom water is thicker in the ½×C ocean. Although it is not shown here, the salinity of deep water is high, particularly in high southern latitudes. The formation of the thick layer of cold and saline deep water described above is attributable mainly to the deep convection that predominates in the Southern Ocean, where sea ice forms rapidly in winter at the oceanic surface, yielding patches of saline and cold water through brine rejection. Although a similar process operates in the 1×C experiment, as described in chapter 8, the rate of deep water formation is substantially smaller.

The ½×C ocean is characterized not only by the thick layer of cold and saline deep water, but also by very extensive and thick perennial sea ice that extends to ~50° S in the Southern Ocean (figure 9.5). This is in contrast to the control, in which the coverage of sea ice undergoes large seasonal variation with little sea ice in summer (see figure 8.9). Albedo feedback plays an important role in maintaining this extensive sea ice cover, but an additional factor of importance is the upwelling of cold deep water driven by the intense westerly winds, which are much stronger than in the 1×C run. As this cold deep water upwells and reaches the surface, it freezes rapidly under the influence of the frigid overlying air. Brine rejection produces cold saline water that induces deep convective mixing. The combination of the upwelling of cold deep water and deep convective mixing prevents the poleward retreat of sea ice in spring and is responsible for the development of thick and extensive sea ice cover. In short, the Southern Ocean of the ½×C simulation may be characterized as a "gigantic sea-ice-producing machine" that also yields cold and saline deep water with near-freezing temperatures in the world's oceans.

In the ½×C simulation, the atmospheric CO_2 concentration is half that of the 1×C simulation. This reduction in CO_2 is substantially larger than the reduction of the CO_2-equivalent greenhouse gas concentration at the LGM relative to its preindustrial value, as indicated by the analysis of air bubbles trapped in the Antarctic ice sheet (e.g., Neftel et al., 1982). Despite the larger greenhouse gas forcing in the ½×C simulation, it is likely that

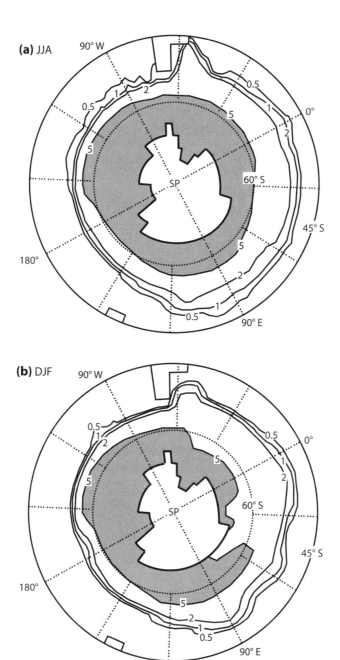

FIGURE 9.5 Geographic distribution of seasonal-mean thickness of sea ice (m) obtained from ½×C, (*a*) for June–July–August (JJA); (*b*) for December–January–February (DJF).

cold and saline deep water with near-freezing temperatures also occupied the deep ocean at the LGM, as indicated by the isotopic and chemical analysis of pore water conducted by Schrag et al. (2002) and Adkins et al. (2002). This suggests that a mechanism of deep water formation similar to that in the ½×C simulation may have operated at the LGM. According to the analysis of deep sea sediments conducted by Cooke and Hays (1982), sea ice covered the Southern Ocean at the LGM during much of the year, with the possible exception of summer (Crosta et al., 1998). One can therefore speculate that very extensive thick sea ice capped the Southern Ocean at the LGM, severely limiting the sea-to-air CO_2 flux in the primary region of deep water ventilation, as suggested by Stephens and Keeling (2000). Owing to sea ice capping at the oceanic surface and enhanced CO_2 solubility in the thick layer of near-freezing temperature, it is likely that deep water dissolved and sequestered huge amounts of carbon, thereby reducing the amount stored in the atmosphere. Meanwhile, the supply of nutrients to the upper layer of the Southern Ocean increased because of the intensification of upwelling of deep water, enhancing the biological production and drawdown of atmospheric CO_2, as suggested, for example, by Shackleton et al. (1983, 1992).

The state of the Southern Ocean obtained from the ½×C simulation resembles the state simulated by Shin et al. (2003) for the LGM, using a coupled atmosphere-ocean model developed by the National Center for Atmospheric Research. Common features of the two simulations include intense westerlies, extensive sea ice, and the formation of cold and saline deep water. The resemblance between these two states of the Southern Ocean suggests that the reduced concentration of atmospheric CO_2 has the dominant impact on the state of the LGM coupled atmosphere-ocean system in the Southern Hemisphere. This is in contrast to the Northern Hemisphere, where continental ice sheets had the dominant impact on the climate at the LGM, as discussed by Broccoli and Manabe (1987).

As noted in chapter 7, Broccoli (2000) made an attempt to simulate the condition of the oceanic surface at the LGM, using an improved version of the atmosphere/mixed-ocean model, in which the heat exchange between the mixed layer and subsurface layer of ocean was prescribed. Although the model simulated well the glacial-interglacial difference in SST reconstructed by CLIMAP at most latitudes, it underestimated the difference in the Southern Ocean poleward of ~40° S, as shown in figure 7.6. It is likely that this discrepancy is at least partially attributable to the absence of the combination of upwelling of cold deep water with near-freezing temperatures and deep convective mixing in the Southern Ocean in his LGM simulation.

In this chapter, we have described a slow but powerful feedback that is likely to play a very important role in the development of glacial climate. Operating in the Southern Ocean, it involves not only the albedo feedback of very extensive sea ice, but also other processes such as the upwelling of cold deep water, rapid freezing of seawater and brine rejection at the oceanic surface, deep convection, and the formation of cold and saline water in the deep ocean. The isotopic and chemical analyses of deep sea sediments suggest that a very thick layer of cold and saline deep water and very extensive sea ice also existed at the LGM, as noted above, and played a critically important role in maintaining the low atmospheric CO_2 concentration and the cold glacial climate.

CHAPTER 10

Global Change in Water Availability

Acceleration of the Water Cycle

Global warming involves changes not only in temperature but also in the rates of evaporation and precipitation. If the concentration of greenhouse gases increases in the atmosphere, the downward flux of longwave radiation increases at the Earth's surface, as explained in chapter 1. Thus, temperature increases at the surface, enhancing evaporation from the surface, as satu-ration vapor pressure increases with increasing temperature according to the Clausius-Clapeyron equation of thermodynamics. The increase in the rate of evaporation in turn results in an increase in the rate of precipitation, thereby accelerating the water cycle of the entire planet.

A first attempt to evaluate the impact of global warming on the water cycle was made by Manabe and Wetherald (1975). The model used for their study was the simple 3-D GCM with highly idealized geography that was described at the beginning of chapter 5. In their numerical experiment, the equilibrium response of climate was obtained as the difference between two long-term integrations of the model with the standard and doubled concentrations of atmospheric CO_2. Toward the end of each run, the global mean rate of precipitation in each of these integrations was identical to that of evaporation, satisfying the water balance at the Earth's surface as well as in the atmosphere. The global mean rates of precipitation (and evap-oration) were 93 cm year^{-1} for the control run and 100 cm year^{-1} for the CO_2-doubling run. This implies that the water cycle of the model increased by about 7.4% in response to the doubling of the atmospheric CO_2 concen-tration. The magnitude of this increase is larger than might be expected, when keeping in mind that the warming resulting from CO_2 doubling was similar to that resulting from a 2% increase in solar irradiance, as noted in chapter 5. Why does the water cycle intensify by 7.4% in a simulation where

TABLE 10.1 *Mean heat budget of the Earth's surface*

	Control	**2×CO$_2$**	**Change (%)**
Downward radiative flux (DSX − ULX)	**102.6**	**106.1**	**+3.5 (+3.4%)**
Net downward solar flux (DSX)	166.0	165.3	−0.7 (−0.4%)
Net upward longwave flux (ULX)	63.5	59.3	−4.2 (−6.6%)
Upward heat flux (LH + SH)	**102.6**	**106.1**	**+3.5 (+3.4%)**
Latent heat flux (LH)	75.4	81.0	+5.6 (+7.4%)
Sensible heat flux (SH)	27.2	25.1	−2.1 (−7.7%)

Heat budget obtained from the simple GCM described in chapter 5, section Polar Amplification. Units are W m^{-2}. From Manabe and Wetherald (1975).

the warming is comparable to one in which the incoming solar radiation increases by only 2%? To try to answer this question, we will examine the heat budget of the Earth's surface.

In the model described here, the Earth's surface has no heat capacity. Thus, its heat balance must be maintained between the net downward fluxes of solar and longwave radiation and the net upward fluxes of sensible heat and latent heat of evaporation, as indicated in table 10.1. This table shows that the net downward flux of radiation increases by 3.4% in response to the doubling of atmospheric CO$_2$, mainly owing to the increase in the downward flux of longwave radiation. On the other hand, the net upward heat flux also increases by an identical amount. Thus, the Earth's surface returns all of the radiative energy it receives back to the overlying troposphere through the upward heat flux. The upward flux of latent heat through evaporation increases by as much as 7.4% (or 5.6 W m^{-2}), whereas the flux of sensible heat decreases by 7.7% (or 2.1 W m^{-2}). Because of the partial compensation between these changes, the upward heat flux increases by 3.4% (or 3.5 W m^{-2}), which is equal to the percentage increase in the net downward flux of radiation, thus maintaining the heat balance at the Earth's surface. In summary, the latent heat of evaporation increases by as much as 7.4%, even though the downward flux of radiation increases by only 3.4%. Why does the latent heat flux increase so disproportionately?

According to the Clausius-Clapeyron equation, the saturation vapor pressure of air increases with increasing temperature at an accelerating pace. This implies that the atmospheric vapor pressure over a saturated surface (e.g., the oceanic surface) also increases nonlinearly as surface temperature increases linearly. Thus, it becomes much easier to remove heat from the Earth's surface through evaporation than through the sensible heat flux as temperature increases. This is an important reason why the rate of evaporation increases by 7.4%, thereby accelerating the pace of the hydrologic cycle, whereas the sensible heat flux decreases by 7.7% (see table 10.1).

As noted in chapter 5, the mean surface temperature of the model increases by 2.9°C in response to the doubling of the atmospheric CO_2 concentration. Given that the mean rates of both evaporation and precipitation increase by 7.4%, this implies that the water cycle intensifies by 2.6% per 1°C increase in mean surface temperature. The hydrologic sensitivity of the model thus obtained may be compared with that of other models constructed more recently. Allen and Ingram (2002) estimated the average hydrologic sensitivity of the models that were used in the IPCC *Third Assessment Report*. They found that the global mean rate of precipitation increases by about 3.4% per 1°C increase in surface temperature. Held and Soden (2006) estimated a hydrologic sensitivity of 2% per 1°C for a subset of the coupled atmosphere-ocean models used in the IPCC *Fourth Assessment Report*. The hydrologic sensitivity of 2.6% per 1°C for the model used by Manabe and Wetherald (1975) happens to lie between the average sensitivities obtained from the two sets of more recent models used for these IPCC assessments.

So far, we have discussed the simulated area-averaged changes in precipitation and evaporation that accompany global warming. As shown in a review of the early studies conducted by Manabe and Wetherald (1985), for example, these changes are not uniform spatially. This is attributable in no small part to changes in the horizontal transport of water vapor by the large-scale circulation of the atmosphere. When tropospheric temperature increases in response to an increase in greenhouse gas concentration, it is expected that the absolute humidity of air would also increase, owing mainly to the increase in saturation vapor pressure (i.e., the moisture-holding capacity of air) with increasing temperature. The increase in absolute humidity in turn enhances the transport of water vapor by the atmospheric circulation. This is an important reason for the changes in the spatial distribution of the difference between precipitation and evaporation due to global warming, which thereby alters the availability of water at the continental surface.

In the remainder of this chapter, we will describe how the distributions of precipitation and evaporation change in response to the doubling and quadrupling of the atmospheric concentration of CO_2, thereby affecting the spatial patterns of river discharge and soil moisture at the continental surface. Our description of these changes will be based upon analysis of the two sets of numerical experiments conducted in the late 1990s using the coupled model described in chapter 8. The first set of numerical experiments simulates the changes in the distributions of precipitation and evaporation and associated changes in water availability (e.g., rate of river discharge and soil moisture) that could occur around the middle of the twenty-first century, when the CO_2-equivalent concentration of greenhouse gases is likely to have doubled. The second set simulates the changes in response to the quadrupling of CO_2-equivalent greenhouse gas concentration. Comparing the results from these two sets of experiments, we will attempt to identify the robust changes that are common to the two sets and elucidate the physical mechanisms that control these changes.

Numerical Experiments

The coupled atmosphere-ocean-land model used for this study consists of GCMs of the atmosphere and ocean and a simple model of the heat and water budgets over the continents. It is similar to the coupled model described in chapter 8, except that the grid size is halved from ~500 to ~250 km in order to better simulate the geographic distribution of precipitation. A water budget of the continental surface is computed at each grid cell for a simple "bucket" that has a globally constant moisture-holding capacity of 15 cm, representing the difference between the field capacity and the wilting point integrated over the root zone of the soil (Manabe, 1969). Evaporation is a function of soil moisture and potential evaporation, which is computed assuming a saturated land surface (Milly, 1992). Where the predicted water content of the bucket exceeds its capacity, any excess water is converted to runoff, which is then collected over river basins and transported to the ocean at each river mouth.

Plate 2a illustrates the geographic distribution of the annual mean precipitation obtained from the control experiment, in which the CO_2 concentration is held fixed at the standard value of 300 ppmv. For comparison, plate 2b illustrates the observed distribution compiled by Legates and Willmott (1990). Inspecting these two panels, one can see that the coupled model simulates reasonably well the large-scale distribution of precipitation. For example, the model places realistically the regions of heavy precipitation in the western tropical Pacific, tropical Africa, and the Amazon basin in South America. It also places well regions of meager precipitation not only over the subtropical oceans, but also in Australia, South Africa, the North American Great Plains, and Central Asia. Further inspection also reveals that the model substantially underestimates precipitation over the tropical oceans, probably owing to the failure of the model to resolve intense tropical storms, which preferentially generate very intense precipitation over the ocean. On the other hand, possibly as a consequence of the oceanic precipitation deficit, the model overestimates precipitation over the tropical continents.

Using this model, Wetherald and Manabe (2002) performed a set of several numerical experiments, in which the model was forced by gradually increasing the CO_2-equivalent concentration of greenhouse gases. The temporal variation of the CO_2-equivalent concentration of greenhouse gases used here follows approximately the IS92a scenario of IPCC (1992) and is shown by a solid line in figure 10.1. As this figure shows, the CO_2 concentration increases at a gradually increasing rate until 1990, after which it increases at the rate of 1% per year (compounded), doubling by the middle of the twenty-first century. The CO_2 growth curve lies in the middle range of the scenarios that were presented in the "Special Report on Emission Scenarios" constructed by the IPCC (2001). The effects of sulfate aerosols, which scatter incoming solar radiation and partially compensate the warming effect of increasing greenhouse gases, were also prescribed, based on Haywood et al. (1997). The sulfate concentration of aerosols used here was estimated and projected for the period 1865–2090—that is, the period of the numerical experiment described here.

The numerical experiment described above was repeated eight times, starting from slightly different initial conditions extracted randomly from the control run, to produce an ensemble of simulations with the same radiative forcing. An ensemble mean for the 30-year period 2035–65 was produced by averaging the model output across the eight ensemble members. This approach greatly reduces the influence of unforced interannual and interdecadal variations, which can have a substantial effect on the hydroclimate. The influence of CO_2 doubling was then estimated as the difference between the 30-year ensemble mean centered on 2050, when the prescribed CO_2 concentration has doubled, and the 100-year mean obtained from the control experiment in which the CO_2 concentration was held fixed at its standard value (1×C).

In addition to the numerical experiments described above, another experiment was conducted (Manabe et al., 2004a) in which CO_2 concentrations were prescribed to increase even further. This experimental design, originally used by Manabe and Stouffer (1993, 1994), was motivated by the work of Walker and Kasting (1992), who conjectured that the atmospheric CO_2 concentration is likely to increase by a factor of three to six in a few centuries unless the combustion of fossil fuels is reduced markedly. In this experiment, the CO_2-equivalent concentration of greenhouse gases increases at a rate of 1% per year (compounded) until it reaches four times

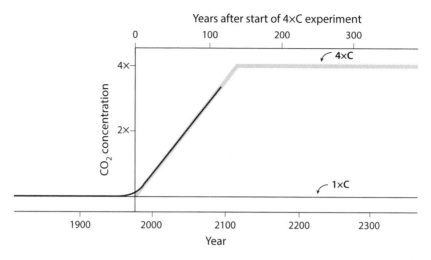

FIGURE 10.1 Time series of CO_2 concentration (logarithmic scale) used for the ensemble of eight experiments with gradually increasing CO_2 and the control (1×C) are indicated by solid black lines. The broad gray line indicates the time series for the CO_2-quadrupling experiment (4×C). On the ordinate, 2× and 4× denotes twice and four times the standard concentration of CO_2 (300 ppmv).

the initial (control) value, and remains unchanged thereafter, as indicated by the broad gray line in figure 10.1. According to this scenario, the CO_2-equivalent concentration would quadruple by the early twenty-second century. The effect of anthropogenic aerosols was not included in this experiment because it is likely to be relatively small on a centennial time scale, as emission controls on sulfur dioxide are strengthened. In order to reduce the effects of unforced variability on interannual and interdecadal time scales, the model output was averaged over the 100-year period between the 200th and 300th years of the experiment. The influence of CO_2 quadrupling was then estimated as the difference between this 100-year mean and a multicentury mean from the control experiment, in which the CO_2 concentration was held fixed at the standard value.

In the eight-member ensemble described above, the global mean surface air temperature increased by about 2.3°C and global mean precipitation increased by 5.2% by the middle of the twenty-first century, when the CO_2 concentration has doubled. In the CO_2-quadrupling experiment, the global mean temperature increased by 5.5°C and global mean precipitation increased by 12.7% a few centuries after the beginning of the experiment, when CO_2 has already quadrupled. In both cases, the hydrologic sensitivity was about 2.3% per 1°C increase in global mean surface temperature, which is similar to the sensitivity of 2.5% per 1°C that Manabe and Wetherald (1975) obtained using the simple model described in chapter 5.

Plate 3a illustrates the geographic distribution of the ensemble mean increase in surface air temperature for the middle of the twenty-first century (i.e., the time of CO_2 doubling), which was obtained from the first set of experiments. Plate 3b illustrates the pattern of surface temperature change obtained from the CO_2-quadrupling experiment. Although the latter is more than twice as large as the former, owing mainly to the much larger positive radiative forcing in the CO_2-quadrupling experiment, the two warming patterns are very similar. Both of them resemble quite well the multimodel mean warming pattern from the models used for the IPCC *Fifth Assessment Report* (see figure 12.10 of Collins et al. [2013]). The basic physical processes responsible for the warming pattern were the subject of detailed analysis presented in chapter 8.

Plate 3b indicates that the magnitude of the warming due to CO_2 quadrupling increases with latitude in the Northern Hemisphere and is more than 14°C over the Arctic Ocean. With the exception of the Southern Ocean, where the warming is delayed greatly as explained in chapter 8, it is comparable in magnitude to the difference in surface temperature between the mid-Cretaceous period (approximately 100 million years ago) and the present, which Barron (1983) estimated referring to various proxy signatures of past climate change. It is therefore likely that the water cycle of the mid-Cretaceous may have been as intense as that of the 4×CO_2 world that will be described in the remainder of this chapter.

We have already discussed how the warming of the Earth's surface affects evaporation, which, in turn, affects precipitation. Analyzing the results from the two sets of experiments described above, the large-scale distributions of the changes in evaporation and precipitation obtained from the CO_2-quadrupling experiment were found to be similar to those from the CO_2-doubling experiment, although the magnitude of the former was about twice as large as the latter. In the following section, we show how the latitudinal profile of precipitation and that of evaporation change in the CO_2-quadrupling experiment, exploring the physical mechanisms that control these changes.

The Latitudinal Profile

Figure 10.2 shows the latitudinal distribution of the zonally averaged annual mean rates of precipitation and evaporation simulated by the model. Although both precipitation and evaporation tend to be larger at low latitudes than at high latitudes, their profiles are different. For example, the zonal mean rate of precipitation is larger than that of evaporation in the

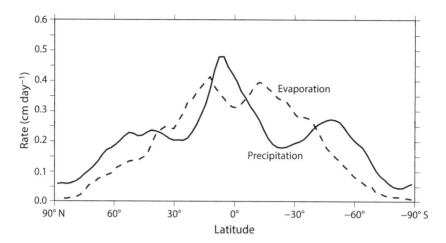

FIGURE 10.2 Latitudinal profiles of the zonally averaged annual mean rates of evaporation and precipitation from the control run, in which the CO_2-equivalent concentration was held fixed at the standard value. From Wetherald and Manabe (2002).

tropics and from middle to high latitudes, whereas the reverse is the case in the subtropics. The difference between the two profiles is attributable mainly to the meridional transport of water vapor by large-scale circulation in the atmosphere.

An important factor that controls the latitudinal profiles of precipitation and evaporation at low latitudes is the Hadley circulation, with rising motion in the tropics and sinking motion in the subtropics. In the near-surface layer of the atmosphere, the trade winds carry moisture-rich air from the subtropics toward the intertropical convergence zone (ITCZ), where intense upward motion predominates and precipitation is at a maximum. In the subtropics, on the other hand, air sinks over a broad zonal belt. Because of adiabatic compression of this sinking air, relative humidity is low, enhancing evaporation from the oceanic surface. Thus, evaporation is at a maximum in the subtropics.

At middle latitudes, where extratropical cyclones are frequent, precipitation is at a maximum. These cyclones carry warm, humid air poleward and cold, dry air equatorward, yielding a net transport of water vapor from the subtropics toward the middle and high latitudes. Thus, atmospheric circulation transports water vapor from the subtropics, where evaporation exceeds precipitation, to the middle and high latitudes, where precipitation exceeds evaporation.

Changes in the latitudinal profiles of precipitation and evaporation in response to the quadrupling of the atmospheric CO_2 concentration are

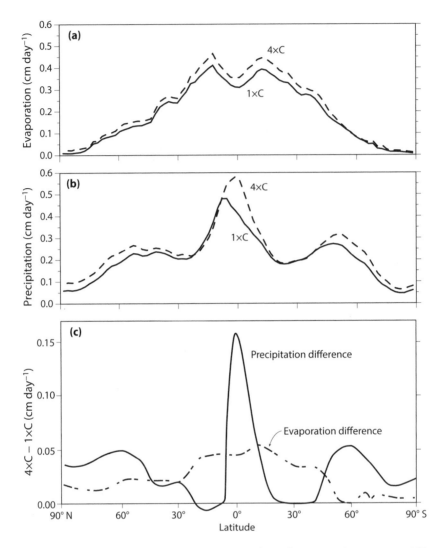

FIGURE 10.3 Latitudinal profiles of (*a*) zonally averaged annual mean rate of evaporation and (*b*) zonally averaged annual mean rate of precipitation, obtained from the control run (1×C) and CO_2-quadrupling run (4×C), and (*c*) differences in evaporation and precipitation between the two runs.

shown in figure 10.3. Figure 10.3a presents the latitudinal profiles of the zonally averaged annual mean evaporation rate from the control run (1×C) and the CO_2-quadrupling run (4×C). A comparison of these two profiles shows that the rate of evaporation increases at all latitudes with the increase in atmospheric CO_2 concentration. The magnitude of the increase is large in the tropics and decreases poleward, becoming small in high latitudes. Because the vapor pressure of air in contact with a wet surface (e.g., the ocean) increases with increasing surface temperature according to the Clausius-Clapeyron equation, the vertical gradient of vapor pressure above the surface also increases as long as the relative humidity of overlying air does not change substantially. This is the main reason why the increase in the rate of evaporation is largest at low latitudes, where the surface temperature is highest, and decreases with increasing latitude.

In response to the increase in evaporation described above, the zonal mean rate of precipitation also increases at most latitudes, as shown in figure 10.3b. The latitudinal distribution of the change in precipitation rate, however, is quite different from that of evaporation. As shown on a magnified scale in figure 10.3c, the rate of precipitation increases much more than that of evaporation in the tropics and in middle and high latitudes, whereas the reverse is the case in the subtropics. These changes are attributable mainly to the increase in the export of moisture from the subtropics toward other latitudes.

When temperature increases in the troposphere owing to global warming, absolute humidity also increases, keeping relative humidity almost unchanged, as discussed earlier. The increase in absolute humidity in turn results in an increase in the transport of tropospheric water vapor. For example, the poleward transport of water vapor by extratropical cyclones increases, thereby increasing the transport of moisture from the subtropics toward middle and high latitudes. Thus, precipitation increases more than evaporation poleward of 45° in both hemispheres, as shown in figure 10.3c. Meanwhile, the equatorward transport of water vapor by the trade winds also increases, increasing the supply of moisture toward the ITCZ, where precipitation increases much more than evaporation. On the other hand, because of the increase in the export of moisture toward both high and low latitudes, precipitation hardly changes in the subtropics, despite an increased supply of moisture through evaporation from the Earth's surface. Held and Soden (2006) found similar hydrologic responses in their analysis of the climate change experiments used for the IPCC *Fourth Assessment Report*. The enhancement of the pattern of precipitation minus evaporation that they identified has been dubbed the "rich-get-richer" mechanism (Chou et al., 2009).

Because of the changes in the rates of precipitation and evaporation described above, water availability at the continental surface changes. For example, the rate of river discharge increases in middle and high latitudes and in the tropics, where precipitation increases more than evaporation. In contrast, over many arid and semiarid regions in the subtropics, soil moisture decreases substantially. Since the downward flux of longwave radiation increases in response to the increase in the concentration of greenhouse gases, the thermal energy available for evaporation increases at the Earth's surface. On the other hand, precipitation hardly increases or decreases in much of the subtropics owing to the increased export of water vapor toward both high and low latitudes. For these reasons, it is expected that soil moisture would decrease substantially over arid and semiarid regions in the subtropics. In the remainder of this chapter, we will describe how the geographic distributions of river discharge and soil moisture change as a consequence of global warming.

River Discharge

When precipitation exceeds evaporation at the continental surface, soil moisture increases. Sooner or later, the soil becomes saturated with water and excess water runs off through rivers. The geographic pattern of annual runoff obtained from the control experiment is presented in plate 4. As expected, the mean annual runoff is usually large in those regions where the rate of precipitation exceeds that of evaporation. For example, the simulated runoff is large in tropical regions of heavy rainfall such as the Amazon basin in South America, the Congo basin in Africa, and river basins in Southeast Asia and the Indonesian islands. The rate of runoff is also large in certain midlatitude regions such as the Saint Lawrence and Columbia basins in North America, and river basins in western Europe. Although the rate of precipitation is not very large, the rate of runoff is large over northern Siberia and northern Canada, where the rate of evaporation is small mainly owing to low surface temperatures under weak solar radiation. On the other hand, runoff is small over many arid and semiarid regions of the continents, such as the Sahara, Central Asia, Great Plains, southwestern North America, much of Australia, and the Kalahari region of Africa, because of meager precipitation and intense incoming solar radiation available for evaporation.

Table 10.2 includes a comparison of historical mean and model-estimated values of annual discharge for a number of important river basins throughout the world. The model estimates were obtained from the control time integration of the model, in which the atmospheric CO_2 concentration is kept unchanged at the standard value throughout the course of the integration. In high and middle latitudes, for example, about half of the basins have modeled discharges within about 20% of the observed value. The total discharge simulated for high and middle latitudes (52,700 and 97,500 $m^{-3} s^{-1}$) compares reasonably well with the observed values (63,200 and 84,400 $m^{-3} s^{-1}$). In low latitudes, however, river discharges are overestimated, particularly in tropical Africa and Southeast Asia. These are also

	River Basin	Mean Discharge (10^3 m^3 s^{-1})[a]		Change (%)[b]	
		Historical	Control	2050	4xC
High Latitude	Yukon	6.5	10.1	+21	+47
	Mackenzie	9.1	8.5	+21	+40
	Yenisei	18.1	12.6	+13	+24
	Lena	16.9	15.1	+12	+26
	Ob'	12.6	6.4	+21	+42
	Subtotal	**63.2**	**52.7**	**+16**	**+34**
Middle Latitude	Rhine/Elbe/Weser/Meuse/Seine	3.9	3.1	+25	+20
	Volga	8.1	5.2	+25	+59
	Danube/Dnieper/Dniester/Bug	8.5	6.7	+21	+9
	Columbia	5.4	6.4	+21	+47
	Saint Lawrence/Ottawa/Saint Maurice/Saguenay/Outardes/Manicouagan	11.8	12.4	+6	+12
	Mississippi/Red	17.9	10.2	+0	−7
	Amur		9.2	−1	+3
	Huang He		16.7	+0	+18
	Chang	28.8	53.5	+4	+28
	Zambezi		31.1	−1	+2
	Paraná/Uruguay		23.5	+24	+54
	Subtotal	**84.4**	**97.5**	**+8**	**+24**
Low Latitude	Amazon/Maicuru/Jari/Tapajos/Xingu	194.3	234.3	+11	+23
	Orinoco	32.9	28.2	+8	+1
	Ganges/Brahmaputra	33.3	48.6	+18	+49
	Congo	40.2	122.3	+2	−1
	Nile	2.8	49.5	−3	−18
	Mekong	9.0	28.6	−6	−6
	Niger		58.3	+5	+6
	Subtotal	**312.5**	**469.8**	**+7**	**+13**
	TOTAL	460.1	661.7	+8	+16

From Manabe et al. (2004b).

[a] Mean discharges shown from historical data and simulated from the control (1×C), where atmospheric CO_2 concentration is kept unchanged at 300 ppmv. Subtotals and totals include only those basins with historical data.

[b] Relative changes simulated to occur from the preindustrial period to the middle of the twenty-first century, when the CO_2 concentration has doubled (2050), and those in response to a quadrupling of atmospheric CO_2 (4×C). The percentage change from *A to B* is defined as $100 \times (B − A)/A$.

regions where precipitation is overestimated substantially (plate 2a and 2b). In general, however, the model reproduces reasonably well the annual discharge from many of the major rivers of the world.

The comparison of observed and modeled discharge presented here is affected by temporal sampling error in the observations, because observational records do not always cover a long enough period to provide precise estimates of climatic means. Additionally, the natural balance between runoff and evaporation is modified significantly as a result of irrigated agriculture and evaporation from artificial reservoirs. Nevertheless, both sampling error and the consequences of water resource development are small compared with the simulation errors in the basins for which comparisons are made in table 10.2.

The geographic distributions of the changes in the annual mean rate of runoff simulated for CO_2 doubling and CO_2 quadrupling are illustrated in plate 5a and b, respectively. As this figure shows, the patterns of the changes are remarkably similar between the two simulations, although the magnitude of the change in the latter case is about twice as large, as one might expect given the difference in the simulated warming. The similarity between the two patterns implies that the physical mechanisms involved are practically identical between the two simulations. It also implies that any effect of unforced variability is very small, largely because of the time averaging applied to the runoff obtained from the two experiments.

In both simulations, runoff increases in high latitudes, particularly over the northwest coast of North America, northern Europe, Siberia, and Canada. It also increases in the rainy regions of the tropics such as Brazil, the west coast of tropical Africa, Indonesia, and northern India. On the other hand, runoff decreases in many semiarid regions such as the zonal belt to the south of the Sahara, the southern part of North America, the west coast of Australia, the Mediterranean coast, and northeast China. The magnitude of the reduction, however, appears to be relatively small in absolute terms, although it may not be small in terms of percentage, as shown by Milly et al. (2008). In general, the geographic pattern of the change in runoff is similar to the multimodel average from the models used for the IPCC *Fifth Assessment Report* (see figure 12.24 of Collins et al. [2013]).

A notable exception occurs, however, in the Amazon basin, where runoff increases substantially in the result presented here but decreases in the multimodel mean. One can speculate that the discrepancy is largely attributable to the difference in the rate of precipitation. As indicated in Flato et al. (2013, fig. 9.4b), the multimodel mean rate of precipitation in the basin is substantially smaller than the observed rate, whereas a similar bias is not evident in plate 2a and 2b for the model presented here. In view of the close relationship between rainfall and runoff, it is likely that the rate of runoff is going to increase in the Amazon basin owing to global warming.

Averaged zonally, changes in the rate of runoff bear some similarity to changes in the difference between precipitation and evaporation. Runoff increases substantially in the tropics and also in middle and high latitudes. In contrast, the magnitude of the change in runoff is small in the subtropics. An increase in the export of water vapor from the subtropics to both higher and lower latitudes is the primary cause of these changes, as discussed earlier in this chapter.

The percentage changes in the rate of river discharge from the major rivers of the world for the CO_2-doubling and CO_2-quadrupling experiments are also shown in table 10.2. As this table indicates, the changes in the quadrupling experiment are about twice as large as in the doubling experiment. For example, the discharge from Arctic rivers such as the Mackenzie and Ob' increases by ~20% with CO_2 doubling and ~40% with CO_2 quadrupling. The large increase in discharge from these Arctic rivers is attributable mainly to the increase in poleward transport of water vapor, as discussed earlier in this chapter. Recently, Peterson et al. (2002) analyzed the time series of the discharges from several major Arctic rivers in Siberia. They found that the total discharge from these rivers has a statistically significant positive trend, in qualitative agreement with the results presented here.

In the middle latitudes, the percentage change in discharge from European rivers such as the Volga is large. The response from these rivers is similar to those in high latitudes. The discharge from Columbia also increases as precipitation increases in the Rocky Mountains. On the other hand, the change in combined discharge of the Paraná and Uruguay rivers is relatively high, reflecting essentially a tropical response within the runoff source region for this system.

In the tropics, the discharge from the Amazon River increases by 11% and 23% in response to the CO_2 doubling and CO_2 quadrupling, respectively. Although the discharges from the Ganges-Brahmaputra and Congo increase greatly in response to the doubling and quadrupling of CO_2, they should be regarded with caution in view of the gross overestimate of the annual discharges obtained from the control experiment. Similar caution may also be applicable to the changes in river discharge from the Congo, Mekong, and Nile rivers.

The impact of climate change on river discharge has also been estimated using stand-alone models of river discharge by Alcamo et al. (1997) and Arnell (1999), and by using climate model output from Vörösmarty et al. (2000) and Arnell (2003). For example, the consensus pattern of change obtained by Arnell (2003) on the basis of data from several climate models is broadly consistent with the pattern described here, with the major exception of the Amazon River, where runoff decreases in his analysis. In sharp contrast, runoff increases substantially in the experiments described here. In view of the substantial underestimation of rainfall in the Amazon basin in many models used in Arnell's analysis, it may be premature, however, to conclude that the discharge from the basin is going to decrease owing to global warming, as noted earlier.

Soil Moisture

A meaningful and direct comparison of modeled soil moisture with observations is not possible because of difficulties in defining the plant-available-water-holding capacity of the soil and because of the extreme heterogeneity of soil moisture, soil properties, and vegetation rooting characteristics. Nevertheless, soil moisture in this model is an excellent indicator of soil wetness. Plate 6 illustrates the distribution of annual mean soil moisture simulated by the model. The model reproduces reasonably well the large-scale features of soil wetness. For example, the regions of very low soil moisture simulated by the model approximately correspond with the major arid regions of the world: the Gobi and Great Indian deserts of Eurasia, the North American deserts, the Australian desert, the Patagonian Desert of South America, and the Sahara and Kalahari deserts of Africa. Furthermore, the model places reasonably well the semiarid regions adjacent to many of the major arid regions in Africa, Australia, and Eurasia. Although the semiarid region in the western plains of North America is simulated by the model, it extends eastward too far, particularly in the southern United States, where precipitation is underestimated substantially (compare plate 2a with 2b). On the other hand, soil moisture is large in Siberia and Canada, located in high northern latitudes where precipitation substantially exceeds the relatively meager evaporation. As expected, soil moisture is also large in heavily precipitating regions of the tropics in South America, Southeast Asia, and Africa. In summary, the model places reasonably well the locations of arid, semiarid, and wet regions of the world.

Drying in Arid and Semiarid Regions

Global warming affects not only river discharge but also soil moisture. The geographic distributions of the change in annual mean soil moisture in response to the doubling and quadrupling of the atmospheric concentration of CO_2 are illustrated in plate 7. The changes are presented in terms of percentage change relative to the control experiment. The geographic pattern of the percentage change in soil moisture due to CO_2 doubling resembles the pattern due to CO_2 quadrupling, though the latter is about twice as large as the former. As noted with regard to the change in the rate of runoff, the similarity between the two patterns implies that the basic physical mechanism involved is practically identical between the two simulations. The percentage reduction in soil moisture is relatively large in many arid and semiarid regions of the world, such as western and southern parts of Australia, southern Africa, southern Europe, northeastern China, and southwestern North America. Although the percentage reduction is also large in the southeastern United States, this result should be regarded with caution because simulated precipitation is substantially less than observed (plate 2) and simulated soil moisture (plate 6) is unrealistically small in the region.

It is encouraging that the geographic patterns of the percentage changes in annual mean soil moisture in plate 7 resemble the multimodel mean pattern of changes from Collins et al. (2013, fig. 12.23) in the IPCC *Fifth Assessment Report*. However, a notable exception occurs in the Amazon basin. Although soil moisture changes only slightly in this basin as indicated in plate 7, it decreases substantially in the multimodel mean. We have previously noted that the multimodel mean precipitation in this basin is much less than observed. A similar discrepancy is not evident in plate 2, which compares the simulated and observed distribution of precipitation for the coupled model discussed here. It is therefore possible that the difference in the sign of soil moisture change may be attributable to the difference in the rate of simulated precipitation in the basin. Thus we are tempted to speculate that soil moisture may increase in the Amazon basin as global warming proceeds, as it does in the present model.

The seasonal dependence of soil moisture change (%) in response to CO_2 quadrupling is shown in plate 8 for each of the standard seasons: June–July–August (JJA), September–October–November (SON), December–January–February (DJF), and March–April–May (MAM). This figure shows that soil moisture decreases in many arid and semiarid regions, particularly during the dry season. For example, the percentage reduction is pronounced in southern Australia from JJA to SON, in and around the Kalahari Desert of Africa in JJA, in southern Europe in JJA, and in southwestern North America from DJF to MAM. Although it is also large in the southeastern United States in MAM, this should be regarded with caution because of the systematic underestimation of precipitation in this region. Although not shown here, the geographic pattern of the soil moisture change that occurs in response to CO_2 doubling resembles the pattern that occurs in response to CO_2 quadrupling.

Why does soil moisture decrease in many arid and semiarid regions of the world? As we have discussed previously, the downward flux of long-wave radiation increases owing to the increase in the concentration of greenhouse gases, thereby increasing the radiative energy that is potentially available for evaporation. On the other hand, the magnitude of the change in precipitation is usually small in these regions. In order to maintain the water balance of the continental surface, it is therefore necessary to reduce evaporation as a fraction of potential evaporation. Because the ratio of evaporation to potential evaporation decreases as the soil dries, a reduction in soil moisture reduces the fraction of the radiative energy used for evaporation. This is the main reason why soil moisture decreases in many arid and semiarid regions of the world. That is not so say that changes in the precipitation rate are not important. Indeed, the percentage reduction of soil moisture tends to be large in those regions where the percentage reduction of precipitation is also large. (For the geographic distribution of the percentage change in precipitation, see Collins et al. [2013, fig. 12.22] from the IPCC *Fifth Assessment Report*.)

In many relatively arid regions of the world, soil moisture is small partly because water vapor is exported outward by the large-scale circulation in the atmosphere, as it is in many regions in the subtropics. Because absolute humidity of air usually increases with increasing temperature, it is expected that the rate of export is likely to increase as global warming proceeds, reducing the amount of water vapor available for precipitation in these regions. This is another reason why soil moisture decreases in such relatively arid regions.

So far, we have discussed the systematic change of soil moisture that occurs on multidecadal to centennial time scales in response to a gradual increase in the atmospheric greenhouse gas concentration. The temporal variation of soil moisture at interannual and decadal time scales is depicted in figure 10.4. The figure illustrates, for both a global warming and a control run, the time series of annual mean and 20-year-running-mean soil moisture over the semiarid region in southwestern North America. The systematic reduction of annual mean soil moisture in the global warming run is often overwhelmed by large natural interannual variability. By the latter half of the twenty-second century, however, the thin gray line signifying the global warming run dips below 3 cm more frequently than the thin black line does that indicates the time series of the annual mean soil moisture obtained from the control run. (In the simple bucket model, a soil moisture value of 3 cm indicates that plant-available water is only 20% of

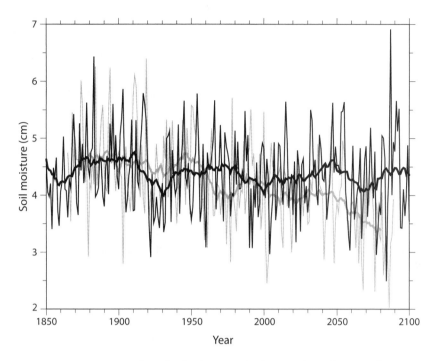

FIGURE 10.4 Time series of simulated soil moisture (cm) averaged over the semiarid region in southwestern North America, which is enclosed by 20° and 38° N latitude, 88° and 114° W longitude, and coastal boundaries. Thin and thick black lines show, respectively, the time series of annual mean and 20-year-running-mean soil moisture obtained from the control integration. Thin and thick gray lines show, respectively, the time series of annual mean and 20-year-running-mean soil moisture obtained from one of the eight-member ensemble of global warming experiments described in the section Numerical Experiments. From Wetherald and Manabe (2002).

what it would be in saturated soil.) This result implies that the frequency of drought is likely to increase during the twenty-first century as plant-available water dips below 20% of saturation in many semiarid and arid regions of the world.

Midcontinental Summer Dryness

As plate 8a shows, soil moisture decreases in summer over extensive mid-continental regions of both North America and Eurasia in the middle and high latitudes. This is in contrast to winter, when soil moisture increases in these regions, as indicated in plate 8c. Midcontinental summer dryness has been the subject of many studies (e.g., Cubasch et al., 2001; Gregory et al., 1997; Manabe and Stouffer, 1980; Manabe and Wetherald, 1985;

Manabe et al., 1992; Mitchell et al., 1990). We will now explore this topic further, referring to the study of Manabe and Wetherald (1987), which was conducted using an atmosphere/mixed-layer-ocean model as described in chapters 5 and 6.

According to their analysis, the large percentage reduction of soil moisture in summer over the northern part of Siberia and Canada around 60° N is attributable mainly to the earlier termination of the snowmelt season. As surface temperature increases at the continental surface owing to global warming, the snowmelt season ends earlier in the spring, exposing the snow-free surface (with low albedo) to intense solar radiation. This is the main reason why the absorption of solar energy at the continental surface increases markedly and makes additional energy available for evaporation in the late spring, thereby reducing soil moisture in summer.

In much of the continental regions in middle latitudes, the summer reduction of soil moisture is attributable not only to the earlier termination of the snowmelt season described above, but also to the poleward shift in the latitudinal profile of precipitation from winter to summer. When temperature increases in the troposphere, the absolute humidity of air usually increases. Thus, the poleward transport of moisture by extratropical cyclones increases, as we have discussed previously. For this reason, the rate of precipitation usually increases substantially along the midlatitude cyclone track and on its poleward flank. In contrast, on the equatorward flank of the cyclone track precipitation hardly changes or decreases slightly, as shown, for example, in figure 10.3b. Because the cyclone track and its associated rain belt shift poleward from winter to summer, particularly over the continents, a midcontinental region located in the poleward flank of the cyclone track in winter would be in its equatorward flank in summer. Thus, precipitation increases substantially in winter, whereas it often decreases slightly in summer. On the other hand, the downward flux of longwave radiation increases at the Earth's surface owing to the increase in atmospheric CO_2 concentration, making additional energy available for evaporation. The combination of increased evaporation and a decrease in precipitation leads to a decrease in soil moisture over midcontinental regions in summer. A similar mechanism also operates in southern Europe, where the percentage reduction in soil moisture is particularly large in summer, as shown in plate 8. There is an important difference, however, because soil moisture in southern Europe decreases not only in summer but also in the other seasons, in contrast to many other regions of middle and high latitudes, where soil moisture increases in winter and early spring. Although precipitation increases in southern Europe in these seasons, the magnitude of the increase is small and is responsible for the

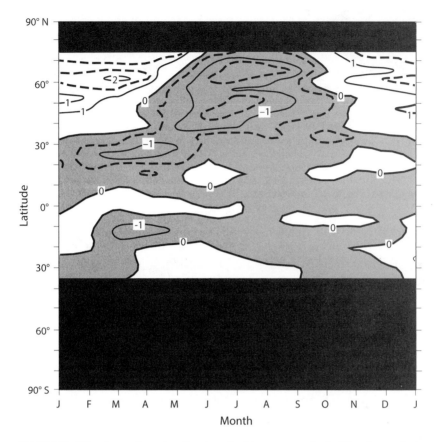

FIGURE 10.5 Zonal mean change in soil moisture (cm) in response to the doubling of atmospheric CO_2 concentration as a function of latitude and calendar month. Black shading indicates latitude bands where there is very little land or the land is ice-covered. From Manabe and Wetherald (1987).

small percentage reduction of soil moisture because it is not large enough to compensate for increased evaporation.

In sharp contrast to summer, soil moisture increases in winter over very extensive regions in both the North American and Eurasian continents in middle and high latitudes, mainly owing to the increase in precipitation. Since temperature is very low in these regions, the rate of evaporation is small and hardly changes in winter despite the increase in surface temperature due to global warming. For these reasons, soil moisture increases over very extensive regions in winter (DJF) and spring (MAM), as shown in plate 8c and d, although it decreases slightly in southern Europe in spring.

Figure 10.5 illustrates the latitude/calendar-month distribution of the equilibrium response of zonal mean soil moisture to the doubling of the atmospheric CO_2 concentration. Although it was obtained from the atmosphere/mixed-layer-ocean model described in chapters 5 and 6, it essentially encapsulates the results that are obtained from the model presented in this section and shown in plate 8. Soil moisture decreases in summer but increases in winter in middle and high latitudes of the Northern Hemisphere. In the subtropics, soil moisture decreases during much of the year, particularly in winter and spring when precipitation decreases. Although the magnitude of the reduction is small in other seasons, it is not necessarily small when expressed in terms of percentage reduction.

Implications for the Future

If the concentration of greenhouse gases continues to increase, following a so-called "business-as-usual" scenario, the reduction of soil moisture in many arid and semiarid regions of the world is likely to become increasingly noticeable during the twenty-first century. By the latter half of the twenty-second century, the reduction of soil moisture in these regions could become very substantial and the frequency of drought is likely to increase markedly. Unfortunately, the river discharge in these regions is not likely to increase significantly, or may actually decrease as global warming proceeds. It is therefore likely that the shortage of water in these regions could become very acute during the next few centuries. In contrast, an increasingly excessive amount of water is likely to be available through river discharge in many water-rich regions in high northern latitudes and in heavily precipitating regions of the tropics, where the frequency of floods is likely to increase markedly, as Milly et al. (2002) found in a numerical experiment. The implied amplification of existing differences in water availability between water-poor and water-rich regions could present a very serious challenge to the water-resources managers of the world. For further discussion of this subject, see the short essay by Milly et al. (2008).

In this book, we have presented, in historical order, many of the studies of global warming that have been conducted since the end of the nineteenth century, when Arrhenius conducted his pioneering study described in chapter 2. These studies have used a hierarchy of climate models with increasing complexity, such as an energy balance model, a 1-D radiative-convective model, and a 3-D GCM of the coupled atmosphere-ocean-land system. These models have been very useful not only for predicting, but also for understanding climate change.

As described in chapter 4, a GCM of the atmosphere consists of prognostic equations of state variables such as wind, temperature, specific humidity, and surface pressure. Each prognostic equation usually consists of two parts. The first is based upon the laws of physics, such as the equation of motion, the thermodynamic equation, Kirchhoff's law of radiative transfer, Planck's function of blackbody radiation, and the Clausius-Clapeyron equation of saturation vapor pressure. The second part includes parameterizations of various sub-grid-scale processes such as moist and dry convection, the formation and disappearance of cloud in the atmosphere, the budget of snow and soil moisture at the continental surface, and the formation and disappearance of sea ice at the oceanic surface, among many others. In the 1960s and 1970s, when early versions of GCMs were developed at various institutions, electronic computers were in the early stages of their development and their capabilities were limited. This is an important reason why the parameterizations of these sub-grid-scale processes in early models were made as simple as possible. Nevertheless, it was encouraging that these models successfully simulated many salient features of the general circulation of the atmosphere and the distributions of temperature and precipitation, as shown, for example, in chapter 4. It is also encouraging that a GCM of the coupled atmosphere-ocean-land system, constructed about 30 years ago, successfully simulated the geographic pattern of surface temperature change that has been observed during the past several decades, as noted, for example, by Stouffer and Manabe (2017).

Owing partly to the simplicity of parameterizations and low resolution, the computational requirements of these models are much smaller than the climate models that are currently used for predicting climate change. Thus, it has been possible to conduct countless numerical experiments, changing one factor at a time, using the models as a virtual laboratory for exploring the inner workings of the climate system. In fact, the simplicity of parameterizations has facilitated greatly the diagnostic analysis of the results obtained. For these reasons, climate models with relatively simple parameterizations have been and are likely to remain very powerful tools for exploring climatic change of not only the industrial present but also the geologic past.

REFERENCES

Adkins, J. F., K. McIntrye, and D. P. Schrag. 2002. "The Salinity, Temperature, and $\delta^{18}O$ of the Glacial Deep Ocean." *Science* 298: 1769–73.

Alcamo, J., P. Döll, F. Kasper, and S. Siebert. 1997. *Global Change and Global Scenario of Water Use and Availability: An Application of Water Gap 1.0.* Kassel, Germany: University of Kassel.

Allen, M. R., and W. J. Ingram. 2002. "Constraints on Future Changes in Climate and Hydrologic Cycle." *Nature* 419: 224–32.

Annan, J. D., and J. C. Hargreaves. 2013. "A New Global Reconstruction of Temperature Changes at the Last Glacial Maximum." *Climate of the Past* 9: 367–76.

Arakawa, A. 1966. "Computational Design for Long-Term Numerical Integration of the Equations of Fluid Motion: Two Dimensional Incompressible Flow." *Journal of Computational Physics* 1: 119–43.

Archer, D., and R. Pierrehumbert, eds. 2011. *The Warming Papers.* Oxford: Wiley-Blackwell.

Arnell, N. W. 1999. "Climatic Changes and Global Water Resources." *Global Environmental Changes* 9: S31–49.

———. 2003. "Effect of IPCC SRES* Emission Scenarios on River Runoff: A Global Perspective." *Hydrology and Earth System Sciences* 7: 619–41.

Arrhenius, S. 1896. "On the Influence of Carbonic Acid in the Air upon the Temperature of the Ground." *London, Edinburgh, and Dublin Philosophical Magazine and Journal of Science*, 5th series, 41: 237–76.

Barkstrom, B. R. 1984. "The Earth Radiation Budget Experiment (ERBE)." *Bulletin of the American Meteorological Society* 65: 1170–85.

Barron, E. J. 1983. "A Warm, Equable Cretaceous: The Nature of the Problem." *Earth-Science Reviews* 19: 305–38.

Beck, J. W., R. L. Edwards, E. Ito, F. W. Taylor, J. Recy, F. Rougerie, P. Joannot, and C. Henin. 1992. "Sea Surface Temperature from Coral Skeletal Strontium-Calcium Ratio." *Science* 257: 644–47.

Berger, A., H. Gallée, T. Fichefet, I. Marsiat, and C. Tricot. 1990. "Testing the Astronomical Theory with a Coupled Climate-Ice Sheet Model." In "Geochemical Variability in the Oceans, Ice and Sediments," edited by L. D. Labeyrie and C. Jeandel, special issue, *Global Planetary Change* 3(1/2): 125–41.

Boucher, O., D. Randall, P. Artaxo, C. Bretherton, G. Feingold, P. Forster, V.-M. Kerminen, et al. 2013. "Clouds and Aerosols." In *Climate Change 2013: The Physical Science Basis. Contribution of Working Group I to the Fifth Assessment Report of the Intergovernmental Panel on Climate Change*, edited by T. F. Stocker, D. Qin, G.-K. Plattner, M. Tignor, S. K. Allen, J. Boschung, A. Nauels, Y. Xia, V. Bex, and P. M. Midgley, 571–657. Cambridge: Cambridge University Press.

Brassell, S. C., G. Eglinton, I. T. Marlowe, U. Pflaumann, and M. Sarnthein. 1986. "Molecular Stratigraphy: A New Tool for Climatic Assessment." *Nature* 320: 129–33.

Broccoli, A. J. 2000. "Tropical Cooling at the Last Glacial Maximum: An Atmosphere–Mixed Layer Ocean Model Simulation." *Journal of Climate* 13: 951–76.

Broccoli, A. J., and S. Manabe. 1987. "The Influence of Continental Ice, Atmospheric CO_2, and Land Albedo on the Climate of the Last Glacial Maximum." *Climate Dynamics* 1: 87–99.

Broccoli, A. J., and E. P. Marciniak. 1996. "Comparing Simulated Glacial Climate and Paleodata: A Reexamination." *Paleoceanography* 11: 3–14.

Broecker, W.S. 1991. "The Great Ocean Conveyor." *Oceanography* 4: 79–89.

Brohan, P., J. J. Kennedy, I. Harris, S.F.B. Tett, and P. D. Jones. 2006. "Uncertainty Estimate in Regional and Global Observed Temperature Change: A New Data Set from 1850." *Journal of Geophysical Research* 111: D12106.

Bryan, K., and M. D. Cox. 1967. "Numerical Investigation of Oceanic General Circulation." *Tellus* 19: 54–80.

Bryan, K., F. G. Komro, S. Manabe, and M. J. Spelman. 1982. "Transient Climate Response to Increasing Atmospheric Carbon Dioxide." *Science* 215: 56–58.

Bryan, K., and L. J. Lewis. 1979. "A Water Mass Model of World Ocean." *Journal of Geophysical Research* 84 (C5): 2503–17.

Bryan, K., S. Manabe, and M. J. Spelman. 1988. "Inter-hemispheric Asymmetry in the Transient Response of a Coupled Ocean-Atmosphere Model to a CO_2 Forcing." *Journal of Physical Oceanography* 18: 851–67.

Budyko, M. I. 1969. "The Effect of Solar Radiation Variations on the Climate of the Earth." *Tellus* 21: 611–19.

Caesar, L., S. Rahmstorf, A. Robinson, G. Feulner, and V. Saba. 2018. "Observed Fingerprint of a Weakening Atlantic Ocean Overturning Circulation." *Nature* 556: 191–96.

Callendar, G. S. 1938. "The Artificial Production of Carbon Dioxide and Its Influence on Temperature." *Quarterly Journal of the Royal Meteorological Society* 64: 223–40.

Cess, R. D., G. L. Potter, J. P. Blanchet, G. J. Boer, A. D. Del Genio, M. Deque, V. Dymnikov, et al. 1990. "Intercomparison and Interpretation of Climate Feedback Processes in 19 Atmospheric General Circulation Models." *Journal of Geophysical Research* 95: 16601–15.

Chapman, W. L., and J. E. Walsh. 1993. "Recent Variation of Sea Ice and Air Temperature in High Latitudes." *Bulletin of the American Meteorological Society* 74: 33–47.

Chappellaz, J., T. Blunier, D. Raynaud, J. M. Barnola, J. Schwander, and B. Stauffer. 1993. "Synchronous Changes in Atmospheric CH_4 and Greenland Climate between 40 and 8 kyr BP." *Nature* 366: 443–45.

Chou, C., J. D. Neelin, C.-A. Chen, and J.-Y. Tu. 2009. "Evaluating the 'Rich-Get-Richer' Mechanism in Tropical Precipitation Change under Global Warming." *Journal of Climate* 22: 1982–2005.

CLIMAP Project members. 1976. "The Surface of the Ice Age Earth." *Science* 191: 1131–36.

———. 1981. *Seasonal Reconstruction of the Earth's Surface at the Last Glacial Maximum.* Map and Chart Series MC-36. Boulder, CO: Geological Society of America.

Collins, M., R. Knutti, J. Arblaster, J.-L. Dufresne, T. Fichefet, P. Friedlingstein, X. Gao, et al. 2013. "Long-Term Climate Change: Projections, Commitments and Irreversibility." In *Climate Change 2013: The Physical Science Basis. Contribution of Working Group I to the Fifth Assessment Report of the Intergovernmental Panel on Climate Change*, edited by T. F. Stocker, D. Qin, G.-K. Plattner, M. Tignor, S. K. Allen, J. Boschung, A. Nauels, Y. Xia, V. Bex, and P. M. Midgley, 1029–136. Cambridge: Cambridge University Press.

Colman, R. 2003. "A Comparison of Climate Feedback in General Circulation Models." *Climate Dynamics* 20: 865–73.

Cooke, D. W., and J. D. Hays. 1982. "Estimates of Antarctic Ocean Seasonal Sea-Ice Cover During Glacial Intervals." In *Antarctic Geoscience*, edited by C. Cradock, 1017–25. Madison: University of Wisconsin Press.

Crosta, X., J.-J. Pichon, and L. H. Burckle. 1998. "Reappraisal of Antarctic Seasonal Sea-Ice at the Last Glacial Maximum." *Geophysical Research Letters* 25: 2703–6.

Crowley, T. J. 2000. "CLIMAP SST Revisited." *Climate Dynamics* 16: 241–25.

Crowley, T. J., and G. H. North. 1991. *Paleoclimatology.* Oxford Monographs on Geology and Geophysics 18. Oxford: Clarendon.

Crutcher, H. L., and J. M. Meserve. 1970. *Selected Level Height, Temperature and Dew Points for the Northern Hemisphere*. NAVAIR 50-IC-52. Washington, DC: US Naval Weather Service.

Cubasch, U., G. A. Meehl, G. J. Boer, R. J. Stouffer, M. Dix, A. Noda, C. A. Senior, S. Raper, K. S. Yap. 2001. "Projection of Future Climate Change." In *Climate Change 2001: The Science of Climate Change*, edited by J. T. Houghton et al., 527–82. Cambridge: Cambridge University Press.

Danabasoglu, G., J. C. McWilliams, and P. R. Gent. 1994. "The Role of Mesoscale Tracer Transports in the Global Circulation." *Science* 264: 1123–26.

Deacon, G.E.R. 1937. "Note on the Dynamics of the Southern Ocean." *Discovery Reports* 15: 125–52.

Deblonde, G., and W. R. Peltier. 1991. "Simulation of Continental Ice Sheet Growth over the Last Glacial-Interglacial Cycle: Experiments with a One-Level Seasonal Energy Balance Model Including Realistic Topography." *Journal of Geophysical Research* 96: 9189–215.

Delworth, T., S. Manabe, and R. J. Stouffer. 1993. "Interdecadal Variations of the Thermohaline Circulation in a Coupled Ocean-Atmosphere Model." *Journal of Climate* 6: 1993–2011.

Denton, G. H., and T. J. Hughes, eds. 1981. *The Last Great Ice Sheets*. New York: John Wiley.

Dixon, K. W., J. L. Bullister, R. H. Gamon, and R. J. Stouffer. 1996. "Examining a Coupled Climate Model Using CFC-11 as an Ocean Tracer." *Geophysical Research Letters* 26: 2749–52.

Edwards, P. N. 2010. *A Vast Machine: Computer Models, Climate Data, and Politics of Global Warming*. Cambridge, MA: MIT Press.

Feigelson, E. M. 1978. "Preliminary Radiation Model of a Cloudy Atmosphere. 1: Structure of Clouds and Solar Radiation." *Contributions to Atmospheric Physics* 51: 203–29.

Flato, G., J. Marotzke, B. Abiodun, P. Braconnot, S. C. Chou, W. Collins, P. Cox, et al. 2013. "Evaluation of Climate Models." In *Climate Change 2013: The Physical Science Basis. Contribution of Working Group I to the Fifth Assessment Report of the Intergovernmental Panel on Climate Change*, edited by T. F. Stocker, D. Qin, G.-K. Plattner, M. Tignor, S. K. Allen, J. Boschung, A. Nauels, Y. Xia, V. Bex, and P. M. Midgley, 741–866. Cambridge: Cambridge University Press.

Forster, P. M. F., and J. M. Gregory. 2006. "The Climate Sensitivity and its Components Diagnosed from Earth Radiation Budget Data." *Journal of Climate* 19: 39–52.

Fourier, J. J. 1827. "Mémoire sur les températures du globe terrestre et des espaces planétaires." *Mémoires de l'Académie royale des sciences de l'institut de France* 7: 569–604.

Fu, Q., and C. M. Johanson. 2005. "Satellite-Derived Vertical Dependence of Tropical Tropospheric Temperature Trends." *Geophysical Research Letters* 32: L10703.

Fu, Q., C. M. Johanson, S. G. Warren, and D. J. Seidel. 2004. "Contribution of Stratospheric Cooling to Satellite-Inferred Tropospheric Temperature Trends." *Nature* 429: 55–58.

Fu, Q., S. Manabe, and C. M. Johanson. 2011. "On the Warming in the Tropical Upper Troposphere: Model versus Observations." *Geophysical Research Letters* 38: L15704.

Gates, W. L. 1976. "Modeling the Ice-Age Climate." *Science* 191: 1138–44.

Gent, P. R., J. Willebrand, T. J. McDougall, and J. C. McWilliams. 1995. "Parameterizing Eddy-Induced Tracer Transport in Ocean Circulation Models." *Journal of Physical Oceanography* 25: 463–74.

Gill, A. E., and K. Bryan. 1971. "Effect of Geometry on the Circulation of a Three Dimensional Southern-Hemisphere Ocean Model." *Deep Sea Research* 18: 685–721.

Goody, R. M. 1964. *Atmospheric Radiation: Theoretical Basis*. Oxford: Clarendon.

Goody, R. M., and Y. M. Yung. 1989. *Atmospheric Radiation: Theoretical Basis*. 2nd ed. Oxford: Oxford University Press.

Gordon, A. L. 1986. "Inter-ocean Exchange of Thermocline Water and Its Influence on Thermohaline Circulation." *Journal of Geophysical Research* 91: 5037–46.

Gregory, J. M., O.J.H. Browne, A. J. Payne, J. K. Ridley, and I. C. Rutt. 2012. "Modelling Large-Scale Ice-Sheet–Climate Interactions following Glacial Inception." *Climate of the Past* 8: 1565–80.

Gregory, J. M., K. W. Dixon, R. J. Stouffer, A. J. Weaver, E. Driesschaert, M. Eby, T. Fichefet, et al. 2005. "A Model Intercomparison of Changes in the Atlantic Thermohaline Circulation in Response to Increasing Atmospheric CO_2 Concentration." *Geophysical Research Letters* 32: L12703.

Gregory, J. M., J.F.B. Mitchell, and A. J. Brady. 1997. "Summer Drought in Northern Midlatitudes in a Time-Dependent CO_2 Climate Experiment." *Journal of Climate* 10: 662–86.

Guilderson, T. P., R. G. Fairbanks, and J. L. Rubenstone. 1994. "Tropical Temperature Variations since 20,000 Years Ago: Modulating Interhemispheric Temperature Change." *Science* 263: 663–65.

Hall, A., and S. Manabe. 1999. "The Role of Water Vapor Feedback in Unperturbed Climate Variability and Global Warming." *Journal of Climate* 12: 2327–46.

Hansen, J., I. Fung, A. Lacis, D. Rind, S. Lebedeff, R. Ruedy, G. Russel, and P. Stone. 1988. "Global Climate Change as Forecast by the Goddard Institute for Space Studies Three Dimensional Model." *Journal of Geophysical Research* 93: 9341–64.

Hansen, J., D. Johnson, A. Lacis, S. Lebedeff, P. Lee, D. Rind, and G. Russell. 1981. "Climate Impact of Increasing Atmospheric Carbon Dioxide." *Science* 213: 957–66.

Hansen J., A. Lacis, D. Rind, G. Russel, P. Stone, I. Fung, R. Ruedy, and J. Lerner. 1984. "Climate Sensitivity: Analysis of Feedback Mechanisms." In *Climate Processes and Climate Sensitivity*, Geophysical monograph 29, Maurice Ewing series 5, edited by J. E. Hansen and T. Takahashi, 130–63. Washington, DC: American Geophysical Union.

Hansen, J., G. Russell, D. Rind, P. Stone, A. Lacis, S. Lebedeff, R. Ruedy, and L. Travis. 1983. "Efficient Three-Dimensional Global Models for Climate Studies: Models I and II." *Monthly Weather Review* 111: 609–62.

Harland, W. B. 1964. "Critical Evidence for a Great Infra-Cambrian Glaciation." *International Journal of Earth Sciences* 54: 45–61.

Harrison, E. F., P. Minnis, B. R. Barkstrom, V. Ramanathan, R. D. Cess, and G. G. Gibson. 1990. "Seasonal Variation of Cloud Radiative Forcing Derived from the Earth Radiation Budget Experiment." *Journal of Geophysical Research* 95: 18687–703.

Hartmann, D. L. 2016. *Global Physical Climatology*. Amsterdam: Elsevier.

Hartmann, D. L., A.M.G. Klein Tank, M. Rusticucci, L. V. Alexander, S. Brönnimann, Y. Charabi, F. J. Dentener, et al. 2013. "Observation: Atmosphere and Surface." In *Climate Change 2013: The Physical Science Basis. Contribution of Working Group I to the Fifth Assessment Report of the Intergovernmental Panel on Climate Change*, edited by T. F. Stocker, D. Qin, G.-K. Plattner, M. Tignor, S. K. Allen, J. Boschung, A. Nauels, Y. Xia, V. Bex, and P. M. Midgley, 159–254. Cambridge: Cambridge University Press.

Hays, J. D., J. Imbrie, and N. J. Shackleton. 1976. "Variations in the Earth's Orbit: Pacemaker of the Ice Ages." *Science* 194: 1121–32.

Haywood, J., R. J. Stouffer, R. J. Wetherald, S. Manabe, and V. Ramaswamy. 1997. "Transient Response of a Coupled Model to Estimated Change in Greenhouse Gas and Sulfate Concentration." *Geophysical Research Letters* 24: 1335–38.

Held, I. M. 1978. "The Tropospheric Lapse Rate and Climate Sensitivity: Experiments with a Two-Level Atmospheric Model." *Journal of Atmospheric Sciences* 35: 2083–98.

———. 1993. "Large-Scale Dynamics and Global Warming." *Bulletin of the American Meteorological Society* 74: 228–41.

Held, I. M., D. I. Linder, and M. J. Suarez. 1981. "Albedo Feedback, the Meridional Structure of the Effective Heat Diffusivity, and Climatic Sensitivity: Results from Dynamic and Diffusive Models." *Journal of the Atmospheric Sciences* 38: 1911–27.

Held, I. M., and B. J. Soden. 2000. "Water Vapor Feedback and Global Warming." *Annual Review of Energy and the Environment* 25: 441–75.

———. 2006. "Robust Response of Hydrologic Cycle to Global Warming." *Journal of Climate* 19: 5686–99.

Held, I. M., and M. J. Suarez. 1974. "Simple Albedo Feedback Models of the Ice Caps." *Tellus* 38: 1911–27.

Henning, C. C., and G. K. Vallis. 2005. "The Effect of Mesoscale Eddies on the Stratification and Transport of an Ocean with a Circumpolar Channel." *Journal of Physical Oceanography* 35: 880–96.

Hoffert, M. I., A. J. Callegari, and C. T. Hsieh. 1980. "The Role of Deep Sea Heat Storage in the Secular Response to Climatic Forcing." *Journal of Geophysical Research* 85: 6667–79.

Hoffman, P. F., A. J. Kaufman, G. P. Halverson, and G. P. Schrag. 1998. "A Neoproterozoic Snowball Earth." *Science* 281: 1342–46.

Holloway, J. L., Jr., and S. Manabe. 1971. "Simulation of Climate by a General Circulation Model. I: Hydrologic Cycle and Heat Balance." *Monthly Weather Review* 99: 335–70.

Hulbert, E. O. 1931. "The Temperature of the Lower Atmosphere of the Earth." *Physical Review* 38: 1876–90.

Imbrie, J., and J. Z. Imbrie. 1980. "Modeling the Climatic Response to Orbital Variations." *Science* 207: 943–53.

Imbrie, J., and K. P. Imbrie. 1979. *Ice Ages: Solving the Mystery*. Hillside, NJ: Enslow.

Imbrie, J., and N. G. Kipp. 1971. "A New Micropaleontological Method for Quantitative Paleoclimatology: Application to a Late Pleistocene Caribbean Core." In *The Late Cenozoic Glacial Ages*, edited by K. K. Turekian, 71–79. New Haven, CT: Yale University Press.

Inamdar, A. K., and V. Ramanathan. 1998. "Tropical and global Scale Interaction among Water Vapor, Atmospheric Greenhouse Effect and Surface Temperature." *Journal of Geophysical Research* 103: 32177–94.

IPCC (Intergovernmental Panel on Climate Change). 1992. *Climate Change 1992: The Supplementary Report to the IPCC Scientific Assessment*. Edited by J. T. Houghton, B. A. Callander, and S. K. Varney. Cambridge: Cambridge University Press.

———. 2001. *Climate Change 2001: The Scientific Basis*. Edited by J. T. Houghton Y. Ding, D. J. Griggs, M. Noguer, P. J. van der Linden, X. Dai, K. Maskell, and C. A. Johnson. Cambridge: Cambridge University Press.

———. 2007. "Acronyms." In *Climate Change 2007: The Physical Science Basis. Contribution of Working Group I to the Fourth Assessment Report of the Intergovernmental Panel on Climate Change*, edited by S. Solomon, D. Qin, M. Manning, Z. Chen, M. Marquis, K. B. Averyt, M. Tignor, and H. L. Miller, 981–87. Cambridge: Cambridge University Press.

———. 2013a. "Acronyms." In *Climate Change 2013: The Physical Science Basis. Contribution of Working Group I to the Fifth Assessment Report of the Intergovernmental Panel on Climate Change*, edited by T. F. Stocker, D. Qin, G.-K. Plattner, M. Tignor, S. K. Allen, J. Boschung, A. Nauels, Y. Xia, V. Bex, and P. M. Midgley, 1467–75. Cambridge: Cambridge University Press.

———. 2013b. "Summary for Policymakers." In *Climate Change 2013: The Physical Science Basis. Contribution of Working Group I to the Fifth Assessment Report of the Intergovernmental Panel on Climate Change*, edited by T. F. Stocker, D. Qin, G.-K. Plattner, M. Tignor, S. K. Allen, J. Boschung, A. Nauels, Y. Xia, V. Bex, and P. M. Midgley, 3–29. Cambridge: Cambridge University Press.

IPCC/TEAP (Technology and Economic Assessment Panel). 2005. *Special Report on Safeguarding the Ozone Layer and the Global Climate System: Issues Related to Hydrofluorocarbons and Perfluorocarbons*. Edited by B. Metz, L. Kuijpers, S. Solomon, S. O. Andersen, O. Davidson, J. Pons, D. de Jager, T. Kestin, M Manning, and L. Meyer. Cambridge: Cambridge University Press.

Jansen, E., J. Overpeck, K. R. Briffa, J.-C. Duplessy, F. Joos, V. Masson-Delmotte, D. Olago, et al. 2007. "Palaeoclimate." In *Climate Change 2007: The Physical Science Basis. Contribution of Working Group I to the Fourth Assessment Report of the Intergovernmental Panel on Climate Change*, edited by S. Solomon, D. Qin, M. Manning, Z. Chen, M. Marquis, K. B. Averyt, M. Tignor, and H. L. Miller, 433–97. Cambridge: Cambridge University Press.

Kaplan, L. D. 1960. "The Influence of Carbon Dioxide Variation on the Atmospheric Heat Balance." *Tellus* 12: 204–8.

Karl, T. R., S. J. Hassol, C. D. Miller, and W. L. Murray, eds. 2006. *Temperature Trends in the Lower Atmosphere: Steps for Understanding and Reconciling Differences*. Washington, DC: US Climate Change Science Program.

Karsten, R. H., and J. Marshall. 2002. "Constructing the Residual Circulation of the ACC from Observation." *Journal of Physical Oceanography* 32: 3315–27.

Kasahara, A., and W. M. Washington. 1967. "NCAR Global General Circulation Model of the Atmosphere." *Monthly Weather Review* 95: 389–402.

Kelly, P. M., P. D. Jones, P. D. Sear, B.S.G. Cherry, and R. K. Tavacol. 1982. "Variation in Surface Air Temperature. 2: Arctic Regions, 1881–1980." *Monthly Weather Review* 110: 71–83.

Klein, S. A., A. Hall, J. R. Norris, and R. Pincus. 2017. "Low-Cloud Feedbacks from Cloud-Controlling Factors: A Review." *Surveys in Geophysics* 38: 1307–29.

Kondratiev, K. Y., and H. I. Niilisk. 1960. "On the Question of Carbon Dioxide Heat Radiation in the Atmosphere." *Pure and Applied Geophysics* 46: 216–30.

Langley, S. P. 1889. "The Temperature of the Moon." *Memoirs of the National Academy of Sciences* 4 (2): 105–212.

Lea, D. W. 2004. "The 100,000-yr Cycle in Tropical SST, Greenhouse Forcing, and Climate Sensitivity." *Journal of Climate* 17: 2170–79.

Legates, D. R., and C. J. Willmott. 1990. "Mean Seasonal and Spatial Variability in Gauge-Corrected Global Precipitation." *International Journal of Climatology* 10: 111–27.

Leith, C. E. 1965. "Numerical Simulation of the Earth's Atmosphere." In *Methods in Computational Physics* vol. 4, edited by B. Alder, S. Fernbach, and M. Rotenberg, 1–28. New York: Academic Press.

Levitus, S. 1982. *Climatological Atlas of the World Ocean.* NOAA Professional Paper 13. Washington, DC: US Department of Commerce.

Levitus, S., J. L. Antonov, T. P. Boyer, and C. Stephens. 2000. "Warming of the World Ocean." *Science* 287: 2225–29.

Loeb, N. G., et al. 2009. "Toward Optimal Choice of the Earth's Top-of-Atmosphere Radiation Budget." *Journal of Climate* 22: 748–66.

London, J. 1957. *A Study of the Atmospheric Heat Balance.* Final Report on Contract AF 19 (122)-165 (AFCRC-TR-57-287). New York: New York University.

Manabe, S. 1969. "Climate and Ocean Circulation. 1: The Atmospheric Circulation and Hydrology of the Earth's Surface." *Monthly Weather Review* 97: 739–74.

Manabe, S., and A. J. Broccoli. 1985. "A Comparison of Climate Model Sensitivity with Data from the Last Glacial Maximum." *Journal of Atmospheric Sciences* 42: 2643–51.

Manabe, S., and K. Bryan. 1969. "Climate Calculation with a Combined Ocean-Atmosphere Model." *Journal of Atmospheric Sciences* 26: 786–89.

Manabe, S., K. Bryan, and M. J. Spelman. 1979. "A Global Ocean-Atmosphere Climate Model with Seasonal Variation for Future Studies of Climate Sensitivity." *Dynamics of Atmospheres and Oceans* 3: 393–426.

Manabe, S., D. G. Hahn, and J. L. Holloway Jr. 1974. "The Seasonal Variation of Tropical Circulation as Simulated by a Global Model of the Atmosphere." *Journal of Atmospheric Sciences* 31: 43–48.

Manabe, S., and J. L. Holloway Jr. 1975. "The Seasonal Variation of the Hydrologic Cycle as Simulated by a Global Model of the Atmosphere." *Journal of Geophysical Research* 80: 1617–49.

Manabe, S., J. L. Holloway Jr., and H. M. Stone. 1970. "Tropical Circulation in a Time Integration of a Global Model of the Atmosphere." *Journal of Atmospheric Sciences* 27: 580–613.

Manabe, S., P.C.D. Milly, and R. T. Wetherald. 2004b. "Simulated Long-Term Changes in River Discharge and Soil Moisture Due to Global Warming." *Hydrological Sciences Journal* 49: 625–42.

Manabe, S., J. Ploshay, and N.-C. Lau. 2011. "Seasonal Variation of Surface Temperature Change during the Last Several Decades." *Journal of Climate* 24: 3817–21.

Manabe, S., J. Smagorinsky, and R. F. Strickler. 1965. "Simulated Climatology of a General Circulation Model with a Hydrologic Cycle." *Monthly Weather Review* 93: 769–98.

Manabe, S., M. J. Spelman, and R. J. Stouffer. 1992. "Transient Response of a Coupled Ocean Atmosphere Model to Gradual Changes of Atmospheric CO_2. Part II: Seasonal Response." *Journal of Climate* 5: 105–26.

Manabe, S., and R. J. Stouffer. 1979. "A CO_2 Climate Sensitivity Study with a Mathematical Model of Global Climate." *Nature* 282: 491–93.

———. 1980. "Sensitivity of a Global Climate Model to an Increase in CO_2 Concentration in the Atmosphere." *Journal of Geophysical Research* 85: 5529–54.

———. 1988. "Two Stable Equilibria of Coupled Ocean-Atmosphere Model." *Journal of Climate* 1: 841–66.

———. 1993. "Century-Scale Effects of Increased Atmospheric CO_2 on the Ocean-Atmosphere System." *Nature* 364: 215–18.

———. 1994. "Multiple-Century Response of a Coupled Ocean-Atmosphere Model to an Increase of Atmospheric Carbon Dioxide." *Journal of Climate* 7: 5–23.

———. 1997. "Coupled Ocean-Atmosphere Model Response to Freshwater Input: Comparison to Younger Dryas Event." *Paleoceanography* 12: 321–36.

———. 1999. "The Role of Thermohaline Circulation in Climate." *Tellus* 51 (A/B): 91–109.

Manabe, S., R. J. Stouffer, M. J. Spelman, and K. Bryan. 1991. "Transient Response of a Coupled Ocean Atmosphere Model to Gradual Changes of Atmospheric CO_2. Part I: Annual Mean Response." *Journal of Climate* 4: 785–818.

Manabe, S., and R. F. Strickler. 1964. "Thermal Equilibrium of the Atmosphere with Convective Adjustment." *Journal of Atmospheric Sciences* 21: 361–85.

Manabe, S., and R. T. Wetherald. 1967. "Thermal Equilibrium of the Atmosphere with a Given Distribution of Relative Humidity." *Journal of Atmospheric Sciences* 24: 241–59.

———. 1975. "The Effect of Doubling CO_2 Concentration on the Climate of a General Circulation Model." *Journal of Atmospheric Sciences* 32: 3–15.

———. 1985. "CO_2 and Hydrology." In *Advances in Geophysics*, vol. 28, *Issues in Atmospheric and Oceanic Modeling*, pt. A, *Climate Dynamics*, edited by S. Manabe, 131–57. New York: Academic Press.

———. 1987. "Large-Scale Changes of Soil Wetness Induced by an Increase in Atmospheric Carbon Dioxide." *Journal of Atmospheric Sciences* 44: 1211–35.

Manabe, S., R. T. Wetherald, P.C.D. Milly, T. L. Delworth, and R. J. Stouffer. 2004a. "Century-Scale Change in Water Availability: CO_2-Quadrupling Experiment." *Climatic Change* 64: 59–76.

Manganello, J., and B. Huang. 2009. "The Influence of Systematic Errors in the Southeast Pacific on ENSO Variability and Prediction in a Coupled GCM." *Climate Dynamics* 32: 1015–34.

Mann, M. E., Z. Zhang, S. Rutherford, R. S. Bradley, M. K. Hughes, D. Shindell, C. Ammann, G. Falvegi, and F. Ni. 2009. "Global Signature and Dynamical Origins of the Little Ice Age and Medieval Climate Anomaly." *Science* 326: 1256–60.

Mann, M. E., Z. Zhang, S. Rutherford, M. K. Hughes, R. S. Bradley, S. K. Miller, S. Rutherford, and F. Ni. 2008. "Proxy-Based Reconstruction of Hemispheric and Global Surface Temperature Variation over the Past Two Millennia." *Proceedings of the National Academy of Sciences of the USA* 105: 13252–57.

MARGO Project members. 2009. "Constraints on the Magnitude and Patterns of Ocean Cooling at the Last Glacial Maximum." *Nature Geoscience* 2: 127–32.

Mastenbrook, H. J. 1963. "Frost-Point Hygrometer Measurement in the Stratosphere and the Problem of Moisture Contamination." In *Humidity and Moisture*, edited by A. Wexler and W. A. Wildhack, vol. 2, 480–85. New York: Reinhold.

Milly, P.C.D. 1992. "Potential Evaporation and Soil Moisture in General Circulation Models." *Journal of Climate* 5: 209–26.

Milly, P.C.D., J. Betancourt, M. Falkenmark, R. M. Hirsch, Z. W. Kundzewicz, D. P. Lettenmaier, and R. J. Stouffer. 2008. "Stationarity Is Dead: Whither Water Management?" *Science* 319: 573–74.

Milly, P.C.D., R. T. Wetherald, K. A. Dunne, and T. L. Delworth. 2002. "Increasing Risk of Great Floods in Changing Climate." *Nature* 415: 514–17.

Mintz, Y. 1965. "Very Long-Term Global Integration of the Primitive Equation of Atmospheric Motion." In *Proceedings of the WMO-IUGG Symposium on Research and Development: Aspects of Long-range Forecasting, Boulder, CO, 1964*, WMO Technical Note 66, 141–67. Geneva: World Meteorological Organization.

———. 1968. "Very Long-Term Global Integration of the Primitive Equation of Atmospheric Motion: An Experiment in Climate Simulation." *Meteorological Monographs* 8 (30): 20–36.

Mitchell, J.F.B., S. Manabe, V. Meleshiko, and T. Tokioka. 1990. "Equilibrium Climate Change and Its Implications for the Future." *Climate Change: The IPCC Scientific Assessment*, edited by J. T. Houghton, G. T. Jenkins, and J. J. Ephrams, 131–72. Cambridge: Cambridge University Press.

Möller, F. 1963. "On the Influence of Changes in the CO_2 Concentration in Air on the Radiation Balance of Earth's Surface and Climate." *Journal of Geophysical Research* 68: 3877–86.

Morice, C. P., J. J. Kennedy, N. A. Rayner, and P. D. Jones. 2012. "Quantifying Uncertainties in Global and Regional Temperature Change Using an Ensemble: The HadCRUT4 Data Set." *Journal of Geophysical Research* 117: D0810.

Morrison, A. K., and A. M. Hogg. 2013. "On the Relationship between Southern Ocean Overturning and ACC Transport." *Journal Physical Oceanography* 43: 140–48.

Munk, W. H. 1966. "Abyssal Recipes." *Deep Sea Research* 13: 707–36.

Neftel, A., H. Oeschger, J. Schwander, B. Stauffer, and R. Zumbrunn. 1982. "Ice Core Sample Measurements Give Atmospheric CO_2 Content during the Past 40,000 Years." *Nature* 295: 220–23.

Newell, R. G., and T. G. Dopplick. 1979. "Questions Concerning the Possible Influence of Anthropogenic CO_2 on Atmospheric Temperature." *Journal of Applied Meteorology* 18: 822–25.

North, G. R. 1975a. "Theory of Energy Balance Climate Models." *Journal of the Atmospheric Sciences* 32: 2033–43.

———. 1975b. "Analytical Solution to a Simple Climate Model with Diffusive Heat Transport." *Journal of the Atmospheric Sciences* 32: 1301–7.

———. 1981. "Energy Balance Climate Models." *Review of Geophysics and Space Physics* 19: 91–121.

PALAEOSENS Project members. 2012. "Making Sense of Palaeoclimate Sensitivity." *Nature* 491: 683–91.

Peixoto, J. P., and A. H. Oort. 1992. *Physics of Climate.* New York: American Institute of Physics.

Perovich, D., R. Kwok, W. Meier, S. Nghiem, and J. Richter-Menge, 2010. "Sea Ice Cover." In "State of the Climate in 2009," edited by D. S. Arndt, M. O. Baringer, and M. R. Johnson. Special Supplement. *Bulletin of the American Meteorological Society* 91: S113–14.

Peterson, B. J., R. M. Holmes, J. W. McClelland, C. J. Vörösmarty, R. J. Lammers, A. I. Shikolomanov, I. A. Shikolamanov, and S. Rahmstorf. 2002. "Increasing River Discharge to the Arctic Ocean." *Science* 298: 2171–73.

Phillips, N. A. 1956. "The General Circulation Model of the Atmosphere: A Numerical Experiment." *Quarterly Journal of the Royal Meteorological Society* 82: 123–64.

Pierrehumbert, R. T. 2004a. "Warming the World." *Nature* 432: 677.

———. 2004b. "Translation of 'Mémoire sur les températures du globe terrestre et des espaces planétaires' by J-B J. Fourier." *Nature* 432 (online supplementary material to Pierrehumbert [2004a]).

Pierrehumbert, R. T., D. S. Abbot, A. Voigt, and D. Koll. 2011. "Climate of the Neoproterozoic." *Annual Review of Earth and Planetary Sciences* 39: 417–60.

Plass, G. N. 1956. "The Carbon Dioxide Theory of Climatic Changes." *Tellus* 8: 140–54.

Po-Chedley, S., and Q. Fu. 2012. "Discrepancies in Tropical Upper Tropospheric Warming between Atmospheric Circulation Models and Satellites." *Environmental Research Letters* 7: 044018.

Pollard, D. 1978. "An Investigation of the Astronomical Theory of the Ice Ages Using a Simple Climate-Ice Sheet Model." *Nature* 272: 233–35.

———. 1984. "A Simple Ice Sheet Model Yields Realistic 100 kyr Glacial Cycles." *Nature* 296: 334–38.

Ramanathan, V. 1975. "Greenhouse Effect Due to Chloro-fluoro-carbons: Climatic Implications." *Science* 190: 50–52.

Ramanathan, V., R. D. Cess, E. F. Harrison, P. Minnis, B. R. Barkstrom, E. Ahmad, and D. Hartmann. 1989. "Cloud-Radiative Forcing and Climate: Results from the Earth Radiation Budget Experiment." *Science* 243: 57–63.

Ramanathan, V., R. J. Cicerone, H. G. Singh, and J. T. Kiehl. 1985. "Trace Gas Trends and Their Potential Role in Climate Change." *Journal of Geophysical Research* 90: 5547–66.

Ramanathan, V., and J. A. Coakley Jr. 1978. "Climate Modeling through Radiative, Convective Models." *Review of Geophysics and Space Physics* 16: 465–89.

Ramanathan, V., and A. M. Vogelmann. 1997. "Greenhouse Effect, Atmospheric Solar Absorption and the Earth's Radiation Budget: From the Arrhenius-Langley Era to the 1990s." *Ambio* 24: 39–46.

Ramaswamy, V., M. D. Schwarzkopf, W. J. Randel, B. D. Santer, B. J. Soden, and G. L. Stenchikov. 2006. "Anthropogenic and Natural Influences in the Evolution of Lower Stratospheric Cooling." *Science* 311: 1138–41.

Riehl, H., and J. S. Malkus. 1958. "On the Heat Balance in the Equatorial Trough Zone." *Geophysica* 6: 503–38.

Schimel, D., I. G. Enting, M. Heimann, T.M.L. Wigley, D. Raynaud, D. Alves, and U. Siegenthaler. 1995. "CO_2 and the carbon cycle." In *Climate Change 1994: Radiative Forcing of Climate Change and an Evaluation of the IPCC IS92 Emission Scenarios*, edited by J. T. Houghton, L. G. Meira Filho, J. Bruce, H. Lee, B. A. Callander, E. Haites, N. Harris and K. Maskell, 35–72. Cambridge: Cambridge University Press.

Schneider, S. H., and S. L. Thompson. 1981. "Atmospheric CO_2 and Climate: Importance of Transient Response." *Journal of Geophysical Research* 86: 3135–47.

Schrag, D. P., J. F. Adkins, K. McIntrye, J. L. Alexander, D. A. Hodell, C. D. Charles, and J. F. McManus. 2002. "The Oxygen Isotopic Composition of Sea Water during the Last Glacial Maximum." *Quaternary Science Review* 21: 331–42.

Screen, J. A., and I. Simmonds. 2010. "The Central Role of Diminishing Sea Ice in Recent Arctic Temperature Amplification." *Nature* 464: 1334–37.

Sellers, W. D. 1969. "A Global Climate Model Based on the Energy Balance of the Earth-Atmosphere System." *Journal of Applied Meteorology* 8: 392–400.

Shackleton, N. J., M. A. Hall, J. Line, and S. Cang. 1983. "Carbon Isotope Data in Core V19-30 Confirm Reduced Carbon Dioxide Concentration of the Ice Age Atmosphere." *Nature* 306: 319–22.

Shackleton, N. J., J. Le, A. Mix, and M. A. Hall. 1992. "Carbon Isotope Records from Pacific Surface Waters and Atmospheric Carbon Dioxide." *Quaternary Science Review* 11: 387–400.

Shin, S., Z. Liu, B. Otto-Bliesner, E. Brady, J. Kutsbach, and S. Harrison. 2003. "A NCAR CCSM Simulation of the Climate at the Last Glacial Maximum." *Climate Dynamics* 20: 127–51.

Smagorinsky, J. 1958. "On the Numerical Integration of the Primitive Equation of Motion for Baroclinic Flow in a Closed Region." *Monthly Weather Review* 86: 457–66.

———. 1963. "General Circulation Experiments with the Primitive Equations. 1: The Basic Experiment." *Monthly Weather Review* 91: 99–164.

Smagorinsky, J., S. Manabe, and J. L. Holloway Jr. 1965. "Numerical Results from a Nine-Level General Circulation Model of the Atmosphere." *Monthly Weather Review* 93: 727–68.

Soden, B. J., and I. M. Held. 2006. "An Assessment of Climate Feedback in Coupled Ocean-Atmosphere Models." *Journal of Climate* 19: 3355–60.

Soden, B. J., and G. A. Vecchi. 2011. "The Vertical Distribution of Cloud Feedback in Coupled Ocean-Atmosphere Models." *Geophysical Research Letters* 38: L12704.

Somerville, R.C.J., and L. A. Remer. 1984. "Cloud Optical Thickness Feedback in the CO_2 Climate Problem." *Journal of Geophysical Research* 89: 9668–72.

Stephens, B. B., and R. F. Keeling. 2000. "The Influence of Antarctic Sea Ice on Glacial-Interglacial CO_2 Variations." *Nature* 404: 171–74.

Stone, P. H. 1978. "Baroclinic Adjustment." *Journal of Atmospheric Sciences* 35: 561–71.

Stouffer, R. J., and S. Manabe. 2003. "Equilibrium Response of Thermohaline Circulation to Large Changes in Atmospheric CO_2 Concentration." *Climate Dynamics* 20: 759–73.

———. 2017. "An Assessment of Temperature Pattern Projection Made in 1989." *Nature Climate Change* 7: 163–65.

Stouffer, R. J., S. Manabe, and K. Bryan. 1989. "Interhemispheric Asymmetry in Climate Response to a Gradual Increase of Atmospheric CO_2." *Nature* 342: 660–62.

Taljaad, J. J., H. van Loon, H. C. Crutcher, and R. L. Jenne. 1969. *Climate of Upper Air. I: Southern Hemisphere*. NAVAIR 50-IC-55. Washington, DC: US Naval Weather Service.

Thompson, S. L., and S. H. Schneider. 1979. "A Seasonal Zonal Energy Balance Climate Model with an Interactive Lower Layer." *Journal of Geophysical Research* 84: 2401–14.

Trenberth, K. E., J. T. Fasullo, and J. Kiehl. 2009. "Earth's Global Energy Budget." *Bulletin of the American Meteorological Society* 90: 311–24.

Trenberth, K. E., P. D. Jones, P. Ambenje, R. Bojariu, D. Easterling, A. Klein Tank, D. Parker, et al. 2007. "Observation: Surface and Atmospheric Climate Change." In *Climate Change 2007: The Physical Science Basis. Contribution of Working Group I to the Fourth Assessment Report of the Intergovernmental Panel on Climate Change*, edited by S. Solomon, D. Qin, M. Manning, Z. Chen,

M. Marquis, K. B. Averyt, M. Tignor, and H. L. Miller, 235–336. Cambridge: Cambridge University Press.

Tsushima, Y., and S. Manabe. 2001. "Influence of Cloud Feedback on the Annual Variation of the Global Mean Surface Temperature." *Journal of Geophysical Research* 106: 22635–46.

———. 2013. "Assessment of Radiative Feedback in Climate Models Using Satellite Observation of Annual Flux Variation." *Proceedings of the National Academy of Sciences of the USA* 110: 7568–73.

Tyndall, J. 1859. "Note on the Transmission of Heat through Gaseous Bodies." *Proceedings of the Royal Society of London* 10: 37, 155–58.

———. 1861. "On the Absorption and Radiation of Heat by Gases and Vapors, and on Physical Connexion of Radiation, Absorption, and Conduction." *London, Edinburgh and Dublin Philosophical Magazine and Journal of Science*, 4th series, 22: 169–94, 273–85.

Vaughan, D. G., J. C. Comiso, I. Allison, J. Carrasco, G. Kaser, R. Kwok, P. Mote, et al. 2013. "Observation: Cryosphere." In *Climate Change 2013: The Physical Science Basis. Contribution of Working Group I to the Fifth Assessment Report of the Intergovernmental Panel on Climate Change*, edited by T. F. Stocker, D. Qin, G.-K. Plattner, M. Tignor, S. K. Allen, J. Boschung, A. Nauels, Y. Xia, V. Bex, and P. M. Midgley, 317–82. Cambridge: Cambridge University Press.

Vecchi, G. A., T. Delworth, R. Gudgel, S. Kapnick, A. Rosati, A. T. Wittenberg, F. Zeng, et al. 2014. "On the Seasonal Forecasting of Regional Tropical Cyclone Activity." *Journal of Climate* 27: 7994–8016.

Vörösmarty, C. J., P. Green, J. Salisbury, and R. B. Lammers. 2000. "Global Water Resources: Vulnerability from Climate Change and Population Growth." *Science* 289: 284–88.

Walker, J.C.G., and J. F. Kasting. 1992. "Effect of Fuel and Forest Conservation on Future Levels of Atmospheric Carbon Dioxide." *Paleogeography, Paleoclimatology, and Paleoecology* 97: 151–89.

Wang, W. C., Y. L. Yung, L. Lacis, A. A. Mo, and J. E. Hansen. 1976. "Greenhouse Effect Due to Man-Made Perturbations to Global Climate." *Science* 194: 685–90.

Washington, W. M., and G. A. Meehl. 1989. "Climate Sensitivity Due to Increased CO_2: Experiment with a Coupled Atmosphere and Ocean General Circulation Model." *Climate Dynamics* 4: 1–38.

Washington, W. M., A. J. Semtner Jr., G. A. Meehl, D. J. Knight, and T. A. Meyer. 1980. "A General Circulation Experiment with a Coupled Atmosphere, Ocean, and Sea Ice Model." *Journal of Physical Oceanography* 10: 1887–1908.

Watts, R. G. 1981. "Discussion of 'Questions Concerning the Possible Influence of Anthropogenic CO_2 on Atmospheric Temperature' by R.G. Newell and T.G. Dopplick." *Journal of Applied Meteorology* 19: 494–95.

Webb, T., and D. R. Clark. 1977. "Calibrating Micropaleontological Data in Climatic Terms: A Critical Review." *Annals of the New York Academy of Sciences* 288: 93–118.

Weertman, J. 1964. "Rate of Growth or Shrinkage of Non-equilibrium Ice Sheet." *Journal of Glaciology* 5: 145–58.

———. 1976. "Milankovitch Solar Radiation Variation and Ice Age Ice Sheet Sizes." *Nature* 261: 17–20.

Weiss, R. F., J. L. Bullister, M. J. Warner, F. A. van Woy, and P. K. Salameh. 1990. *Ajax Expedition Chlorofluorocarbon Measurements*. Scripps Institution of Oceanography (SIO) Reference Series 90-6: 190. La Jolla: University of California, San Diego, SIO.

Wetherald, R. T., and S. Manabe. 1975. "The Effect of Changing the Solar Constant on the Climate of a General Circulation Model." *Journal of Atmospheric Sciences* 32: 2044–59.

———. 1980. "Cloud Cover and Climate Sensitivity." *Journal of Atmospheric Sciences* 37: 1485–510.

———. 1981. "Influence of Seasonal Variation upon the Sensitivity of a Model Climate." *Journal of Geophysical Research* 86: 1194–1204.

———. 1986. "An Investigation of Cloud Cover Change in Response to Thermal Forcing." *Climatic Change* 8: 5–23.

———. 1988. "Cloud Feedback Processes in a General Circulation Model." *Journal of Atmospheric Sciences* 45: 1397–415.

———. 2002. "Simulation of Hydrologic Changes Associated with Global Warming." *Journal of Geophysical Research* 107: 4379–93.

Wielicki, B. A., B. R. Barkstrom, E. F. Harrison, R. B. Lee III, G. L. Smith, and J. E. Cooper. 1996. "Cloud and the Earth's Radiant Energy System (CERES): An Earth Observing System Experiment." *Bulletin of the American Meteorological Society* 77: 853–68.

Wigley, T.M.L., C. M. Ammann, B. D. Santer, and S.C.B. Raper. 2005. "Effect of Climate Sensitivity on the Response to Volcanic Forcing." *Journal of Geopgysical Research* 110: D09107.

Williams, J., R. G. Barry, and W. M. Washington. 1974. "Simulation of the Atmospheric Circulation Using the NCAR General Circulation Model with Ice Age Boundary Conditions." *Journal of Applied Meteorology* 13: 305–17.

Winton, M. 2006. "Surface Albedo Feedback Estimates for the AR4 Climate Models." *Journal of Climate* 19: 359–65.

Yamamoto, G. 1952. "On the Radiation Chart." *Science Reports of Tohoku University*, series 5, 4: 9–23.

Yamamoto, G., and T. Sasamori. 1961. "Further Studies on the Absorption by the 15 Micron Carbon Dioxide Bands." *Science Reports of Tohoku University*, series 5, 13: 1–19.

Zipser, E. J. 2003. "Some Views on 'Hot Towers' after 50 Years of Tropical Field Programs and Two Years of TRMM Data." *Meteorological Monographs* 51: 49–58.

Italic page numbers refer to figures in the text.

observational data: and CLIMAP Project, 95–
105; and cloud fraction in upper and lower
troposphere, 87–88, *88*; and cooling of the
stratosphere, 34–35, *35*; and deep ocean
temperature, *116*; and geographical distribution of precipitation, *47*, 149, plate 2;
and glacial–interglacial climate transitions,
94–96; and heat content of the ocean, 108;
latitude-height distribution of zonal mean
temperature, *45*; need for continued observations of top of the atmosphere flux of
outgoing longwave radiation, 92; and pa-
leoclimate, 92, 94–96, 98–105, *99*, *102*, *104*,
142; and polar amplification of warming,
55, 55–56, 67; and river discharge, 156–58,
plate 4; and sea ice extent, 68–69, *69*, 125,
126; and seasonal dependence of warming,
67–68, *68*; and sea surface temperature,
99; and slowdown of Atlantic overturning
circulation, 129; and surface temperature,
2, *55*, 55–56, *63*, 67–68, *68*, 122; and terrestrial radiation, 5, 6; and top of the atmosphere flux of outgoing radiation, 5, 6, 92;
and vertical temperature structure of the
atmosphere, *31*, 31, 91; and volcanic cooling, 92; and warming of the troposphere,
34, *35*, 52; and warming or cooling effect
of cloud cover, 81; and winds, *39*, *49*. *See
also* corals, data from; deep sea sediments;
ice-core records; radiosonde observations;
satellite observations
ocean, 106–36; Atlantic Meridional Overturning Circulation (AMOC), 117–18, 126–30,
128; and carbon sequestration, 144; and
cold climate and deep water formation,
137–45; convection in, 112, 126, 134–36,
142; and delayed response to thermal
forcing, 107–9, 118, 120, 122, 125–27, 136;
eddies, 133–35; Great Ocean Conveyor,
125–30, *127*; heat balance of, 61; heat capacity of, 107; heat content (observations),
108; heat exchange between surface layer
and deeper ocean, 108–9, 112, 119, 126,
134–37; latitude-depth cross section of
ocean temperature (simulated and observed), *116*, *141*, 141–42; mechanisms of
heat transport, 112; ocean heat transport
in GISS model, 82; surface temperature
(*see* sea surface temperature); temporal
variation of deep water temperature, *139*;
thermal inertia of, 106–13, 118, 126, 132;
turbulence in, 112, 130; zonal mean temperature change (GFDL global warming

experiment), 132, *133*. *See also* atmo-
sphere/mixed-layer-ocean models; coupled
atmosphere-ocean-land models; sea ice;
specific oceans
one-dimensional energy balance models,
59–60. *See also* energy balance models
one-dimensional vertical column models, 25–
36, 75. *See also* radiative-convective models
one-layer model (Arrhenius's model), 19
Oort, A. H., 7
ozone, 4, 7, *9*, 17, 29, 35

Pacific Ocean: and intertropical convergence
zone, 48; low pressure centers, 40; precipitation, 46, 149; sea surface temperature at last
glacial maximum, 101; surface winds, 48
PALAEOSENS Project, 105
paleoclimatic data, 92, 94–96, 98–105, *99*, *102*,
104, 142
Peixoto, J. P., 7
Phillips, N. A., 37–38
Planck feedback, 73–76, 83, 84
Planck function, 8, 167
planetary waves, 50, 57
Plass, G. N., 22
Po-Chedley, S., 91
polar amplification of warming, 50–60; absence in the Southern Ocean, 56, 67–68,
136; and albedo effect, 52, 139; and GFDL
model, 52, 61, 113, 120, *121*, 139–40, *140*,
152; and GISS model, 69, 113; observations, *55*, 55–56, 67; and reduction of tropospheric static stability in high latitudes,
52; and similarity of results for CO_2 doubling and solar irradiance experiments, 57
precipitation: and CO_2 quadrupling experiment, 151–56, *153*; and cyclones, 129, 153,
164; exceeding evaporation in the middle
and high latitudes, 153–56, *154*; geographical distribution (GFDL model), 46–47,
47, 149, plate 2; geographical distribution
(observed), *47*, 149, plate 2; increasing
rate due to global warming, 146, 151, 163;
latitudinal profile, 152–56, *153*, *154*, 164;
seasonal variation, 164. *See also* hydrological cycle; river discharge

Q-flux technique, 82, 103, 119

radiative-convective models, 25–36, 72; 1-D
vertical column model, 25–36, 75; and
cloud cover, 30–31; coupled with ocean
model (Hansen et al., 1981), 109–10;